Terje Tvedt is Professor of Geography at the University of Bergen and Professor of Political Science and of Global History at the University of Oslo. He has published extensively on water-related topics and presented three successful television documentaries on water, shown in 150 countries worldwide. His books include *The River Nile in the Age of the British*, *A Journey in the Future of Water*, and he is the Series Editor of the pioneering *History of Water Series*, all published by I.B.Tauris.

WATER AND SOCIETY

CHANGING PERCEPTIONS OF SOCIETAL AND HISTORICAL DEVELOPMENT

Terje Tvedt

I.B. TAURIS

LONDON · NEW YORK

The book was written with financial support from the University of Bergen, the University of Oslo and the Norwegian Research Council.

Published in 2016 by
I.B.Tauris & Co. Ltd
London • New York
www.ibtauris.com

International Library of Human Geography: 30

ISBN: 978 1 78453 079 2
eISBN: 978 0 85773 904 9

A full CIP record for this book is available from the British Library
A full CIP record is available from the Library of Congress

Library of Congress Catalog Card Number: available

Printed and bound by CPI Group (UK) Ltd, Croydon, CR0 4YY

CONTENTS

ACKNOWLEDGEMENTS

This book summarises what I have learnt about the connections and linkages between society and water since I first started to work on the topic 35 years ago. The number of scholars, librarians, water engineers and fellow travellers who in various ways have supported, helped and inspired me is therefore too many to be listed here. The work on the history and historiography of the River Nile could not have been carried out or completed without the help of a large number of librarians and archivists in many countries. Editing the nine-volume series *A History of Water* with contributions from more than 220 scholars from all kinds of disciplines and coming from almost 100 countries gave me the opportunity to learn from many of the best scholars in the world on water-related issues. Preparing for and filming and interviewing for the three TV documentaries on the history and future of water brought me to more than 50 countries with widely different water-society relations, and everywhere engineers, politicians and guides gave invaluable assistance and input. Some of the chapters have been published individually before, and readers and editors have helped me to improve them. I hereby thank them all.

I must especially, however, thank my family since almost every summer vacation for the last 30 years have been 'travelling holidays' – on the search for stories about water in societies. We have journeyed along winding river basins and visited stunning water bodies, confident that when following water we will also enjoy fascinating journeys in cultural history, in art and in technological development. Most often they have liked the trips just as much as myself, for after all: what is more beautiful than water?

This book is dedicated to those who have helped me throughout the years, but also to all the experts, politicians, researchers and ordinary people who will be forced to analyse and reflect upon the past, current and future roles of water in society in the new age we just have entered: the age of water insecurity.

1

THE NEED FOR A PARADIGM SHIFT

This book addresses a major paradox: in spite of the innumerable confluences between society and water, the social significance of water has made surprisingly little impact on our contemporary understanding of human history and development. New discoveries about our planet, as well as developments in society and nature, demand a shift in how we think about the world, a reorientation of social science and historical research. This book, encouraged by a growing interest in the role of water in history and social development among historians, engineers, social scientists, politicians and the public at large, promises to open up radically new fields of social enquiry. It distances itself from powerful and conventional viewpoints on the relationship between nature and society and on how the distinction between the two has been drawn. It shows how a reorientation of the social sciences and historical research can happen, and proposes an approach that will enable us both to ask new and fruitful questions about social and historical issues, and to answer old questions in a more inclusive, non-reductionist way.

The dominant conceptual and theoretical traditions are still fundamentally water-blind in their analyses and understanding of society, history and climate. But it is a blindness that cannot any more be justified by lack of knowledge. When on Christmas Eve 1968 the first picture of Earth from outer space was taken, we could all suddenly see the vast blue oceans covering three-quarters of our planet's surface; the white expanse of the polar ice caps; and the grey vapour-laden cloud systems enveloping the globe. This image made it dramatically clear that our planet is truly the Water Planet,[1] and we could all see, with our own eyes, what none of the founding fathers of the social sciences could have known. This image of Earth and all the societies on it – small dots surrounded by water on the move – illustrated both the centrality and the particularities of the waterscape on Tellus. It was in this unique environment that the human race became the dominant species and that societies were formed and developed. We now know that the hydrosphere, including the cloud systems, contains an estimated 1.5 billion cubic kilometres of water (enough to cover the entire planet beneath it with hundreds upon hundreds of metres of water), that oceans cover about 70 per cent of the planet's surface, and that much of the remainder – which is normally but wrongly classified as 'dry land' – is actually crossed, and made habitable, by thousands of rivers, or is dotted with lakes, underlaid by huge reservoirs of groundwater, or covered by enormous amounts of water in frozen form: the Antarctic Ice Sheet alone covers an area larger than the USA and Mexico combined!

What that image from 1968 so unmistakably shows is that water and hydrological processes are at the very heart of the Earth system. Biologists have long ago shown that of all the requirements of life the need for liquid water is paramount. We know that every seed and embryo begins its life in water, and that wherever water is found it is theoretically possible that something is metabolising. Everybody agrees that water makes life possible, but more challenging, when it comes to understanding society and nature, is reconstructing the human experience: life should be seen in terms of a continuous and complex series of organic reactions and social actions, all of which are accomplished in an aqueous environment.

The more discoveries that are made about water on other planets the clearer it becomes that what is special about our planet is not the presence of water here, but the unique way that water flows across the planet in huge but varying amounts. Without this water in liquid and gaseous form, in the oceans and in the wind, neither soils, bacteria, plants, animals nor human beings would have developed, nor, of course, would civilisations have evolved. The hydrological cycle and its spatial variations are therefore nothing less than a key component in any non-reductionist explanation of broad-scale patterns of evolution itself as well as of the evolutionary diversity of social and civilisational change.

Research has proven beyond doubt that the water that characterises this planet is also the vital component of the Earth's energy and climate machine. Water circulates continuously throughout the system in a solar-powered process. The land part of the hydrological cycle brings the water back to the oceans via streams and rivers, although some of it disappears into the soil and into underground channels and aquifers. The amount of water in the pores of the soil influences the interaction between land and atmosphere, but also vegetation patterns and types of agricultural production all over the world. Evaporation and re-condensation are the primary energy source for atmospheric motion, so water is not just a passenger on passing winds. It creates to a large extent the breeze that transports it across the oceans and the continents; water is thus both the parent and sibling of the winds of the North Sea and of the monsoons of South Asia. Nowadays we also know that this water cycle is more and more influenced by what happens to water as it passes through society and as societies leave their water footprints. The water cycle should therefore now be conceived of as the product of both nature and society, a coupled result of the hydrological cycle and the hydrosocial cycle influencing each other and where historical development implies a hydrosocial rearrangement.

In general, the effects of climate change have always manifested themselves in changes to the hydrological cycle and in how water runs in the landscape. This has been so in the past and will be so in the future. It is thus of great social interest that water acts as the planet's most important solvent by far, continuously transporting all sorts of natural material and societal waste from one place to another. Water is also the planet's most powerful erosive agent. Today's landscapes are largely a legacy of hydrological processes which, in the course of millennia, have shaped the land through weathering, erosion and sedimentation, and that is also why the same landscapes are vulnerable to changes in the water cycle.

The more that scientists study the human body, the more they find out about how absolutely crucial water is for most bodily functions. Human evolutionary success among the billions of other organisms on earth must to a large extent be explained by our unusual ability to exploit and adapt to variable and changing waterscapes. Like amphibians and reptiles, we have evolved from continuous immersion in water, and water is still absolutely crucial for reproduction and life. Life itself can be seen as a journey from watery birth in the womb to a dehydrated death. Between these two points each and every one of us must struggle to maintain his or her precarious water balance. Most of the components of fluid balance are controlled by homeostatic mechanisms that are activated when deficits or excesses of water reach only a few hundred milliliters. These mechanisms respond to the state of body water, whether we are aware of it or not, and thus water is the body's busiest substance. And unlike a diet, which can easily be replaced by another diet because food can be transported over great distances, there is no substitute for water, the transport cost of which can be prohibitive over large distances.[2]

Since people who lose 10 per cent of their body water mass go insane, and die if they lose 20 per cent, all individuals have their history written in water – from Heraclitus, who died because he misunderstood the need for water balance in his body when he tried treating himself by drying himself in the sun,[3] to the philosopher John Locke who only drank water because he thought it healthy,[4] to the anonymous worker who has a pint of beer every afternoon. Mostly we manage this without giving a single thought to the enduring and complex webs of vital relationships that make this possible.[5] Human actions are notwithstanding fundamentally influenced and structured by the requirements of the components of this fluid balance, whether the actors reflect upon it or not. It is intriguing that these repeated acts and all that they require of social organisation, forming and framing humans' daily lives in a multitude of ways as they do, have been theoreticised in social science to such a limited extent.

Of course, it has gradually become more urgent to understand the interconnections between water and social development because of the growing gap between supply and demand for water in many places in the world, and because of the uncertainty about future waterscapes. The phenomenal growth in irrigated agriculture, industry and urbanisation during recent decades, coupled with the devastating consequences of water-borne diseases, have made water control the number one issue in many areas of the world. Indeed, the water issue is one of growing political and ideological importance – as evidenced by the emerging water crisis in different parts of the world, the fact that climate change manifests itself in societies in the form of drought and flooding, and popular notions that pollution and the damming of large river systems are the very symbols of modernity gone astray. Because water is an absolute necessity for all, within this overall context of supply and demand, the water issue has become a global political and ideological battlefield. Some researchers are calling for a Blue Revolution or a new water revolution, seeing current water crises as mirrors of a wrong development path.

The motivation for this book, however, goes beyond current ideological and political battles over water and its meaning. The overall aim is to further our ability to understand social and historical development as such, and the role of water within it. It forwards a methodology that can be employed in contrastive studies, and in both diachronic and synchronic perspectives, but perhaps more than anything it provides an approach for studying societies in the long term, since all societies have a history in relation to water from the time they first emerged until today and as long as they will exist in the future.

BEYOND IDEOLOGY: TOWARDS AN ONTOLOGY OF WATER

This book does not limit itself to the crucial task of criticising the water blindness that exists in history and the social sciences. It proposes, in addition, ways to study water-society interactions in a systematic, comparative way. As a starting point it suggests an ontology of water in line with analytical concepts and approaches that can provide a fruitful means of interpreting society and history.

What makes our understanding of water-society interactions so crucial is that since water has been essential to all people at all times, all societies – without exception – have been forced to adapt to, or control in one way or another, the water that flows across their landscapes. Water is thus universal. At the same time, the way in which water moves across varies from place to place and from time to time, even at the same location. Water is therefore also particularistic. This particular combination of the universal and the particular is the fundamental reason why it is especially fruitful to study water-society issues comparatively. No other issue can be studied across the board both in time and space in the same way. All societies can be studied from the perspective of (a) how they have been affected by the physical waterscape, (b) how they have modified this waterscape and changed themselves and the environment in this process, and (c) how they have thought about water, its cultural meanings and value. The water-society nexus thus provides a rare opportunity for broad and, at the same time, rigorous comparative research of developments both in nature and in society, and in time and in space.

THE 'WATER-SYSTEM APPROACH'

We thus need an approach that recovers water as an autonomous actor in society, always acknowledging that it is located in a particular place and time, but also tied intrinsically to the larger scale and longer time frame in such a way that it inherits from them many of its structural (hydrological, topological, energy) properties. The historical-geographical archaeology of water-society relations should also maintain the autonomy of the social, including the cultural and spatial contexts and distinctions, as well as those related to the management of and thinking about water. There is a demand for an approach that manages to grasp how the water that flows across and on the planet exists independently of the different cultural perceptions of it, but also accepts, as a truism, that water is always being understood through such cultural lenses, be they religious, engineering or political. In order to be able to map and analyse the intricate, historical and spatial relations between societies and water, this approach must abandon both constructivism and positivism.

Only by looking at water in society and nature in this broad, inclusive way, can the role and impact of water be properly analysed and understood, and the actual history of the growing influence of the hydrosocial cycle and rearrangement be reconstructed. Water is eternal in nature and in society, but it is also always changing in nature and society. Water is both creator and destroyer in nature, as well as in society. Expressed in the language of the social sciences it is both a prerequisite for

5

social development and frames what development options are possible at every junction in time and at every place. It exists both as a physical object and as a non-physical entity, and as an instrument of the engineer and an object of God for the believer.

This book argues in favour of a historical-geographical archaeology of waterscapes and water-society relations, and will at the same time engage critically with past discourses on particular spatially bounded water-society issues. It will reconstruct and analyse how such discourses have reflected cultural traditions and interactions with particular waterscapes and how belief systems and knowledge about water have been rooted in history and must be analysed from a spatial, geographical perspective.

Theories and methodologies will be suggested here that aim not to reduce the natural world or the world of water to a blank slate on which only human actions matter, or to reduce different development trajectories to a question that can be explained with social variables only, as if structures and events in the natural world are of no relevance. The book underlines the importance of realising that hydraulic works and designs reflect both the natural and the social world, and that hydraulic calculations should therefore be an interdisciplinary effort. It must be crucial from this same perspective to be able to analyse and reconstruct changes in the hydrological cycle, in river discharges and floods, but also how people have interacted with and sought to control their water resources and how they have been thinking about their waters, all the time concerned with understanding how waterscapes and societies have been coupled and have co-evolved.

What is here called the 'water-system approach' is intended to encourage this kind of broad, inclusive yet still rigorous analysis, and it therefore consists of three different but interconnected analytical 'layers'.

First layer

The first layer is water's natural (physical and chemical) form and behaviour. This layer highlights the hydrological cycle and the natural, regional and local waterscapes, based on the notion that such geographical and climatic factors have affected and still to varying degrees affect issues like the broad patterns of human migration and settlement, the general emergence and locational patterns of agricultural centres, food-producing regimes and cities, the birthplaces and structure of early industrialisation and important aspects of the current globalisation of industrialisation in new countries. A focus on this layer will also enable fruitful research on how the hydrological cycle has contributed, and still contributes, to the evolution of societal diversity and different development trajectories. Within this perspective it becomes essential to reconstruct issues such as seasonal and annual precipitation and evaporation patterns, river discharges and velocity measurements, aquifers and their behavioural characteristics, and energy transport in water – all in order to understand empirically the actual interconnections and relationships between nature and society when it comes to water.

Since water's unusual natural characteristics have a wide variety of implications for society, it is not sufficient to understand hydrology only, or to reconstruct the patterns and history of the local variant of the water cycle. Water is unusual in many respects, and almost all of its exceptions to many of the rules of nature are reflected in the fabric of social life. It has the highest surface tension of all liquids, it can absorb and release heat more than most other substances, it expands instead of contracting when it freezes, the solid form of water floats on the heavier liquid and water changes from liquid to vapour or ice and vice versa in the blink of an eye or over millions of years – all factors that have far-reaching and amazing social implications. Furthermore, the fact that water as a substance is on the move, and in most cases ultimately evaporates due to solar radiation before it returns to the Earth as rain or snow, makes it difficult to appropriate and claim effective ownership of it. The mere existence of water therefore brings into question dominant theories of property and management, theories fundamental to most discussions about society, but too taken-for-granted in current mainstream research. There is an endless number of cases demonstrating the need to explore in more detail water's different characteristics and the social implications, also because it it precisely these natural characteristics that have made it rational for humans to spin webs of significance and meaning around water in ways that no other element can match.

Finally, since the workings of the hydrological cycle established water as both the most common substance on earth and the most unevenly distributed resource on the planet before the birth of societies, one cannot fully understand social diversity, social distinction and conflict without understanding this physical aspect of water and how societies adapt to it. In most regions the precipitation and the rivers have created and shaped the valleys they water and drain, and they have thus determined where people have settled. How the rivers run and where the run-off from precipitation goes reflect complex interactions between precipitation, catchments and topography, and affect energy and nutrient turnover and the storage and processing of organic substrates, again influencing all sorts of social activities.

An analytical focus on the physical, natural aspects of the water-system highlights another very interesting theoretical and empirical aspect of water: it is both exogenous and a part of society at the same time. Water is not like other elements in nature transformed by being 'socialised'. Water is H_2O in nature just as it is H_2O in society: the same water that thunders down gigantic cataracts flows from taps and in toilets, and is trapped behind massive stone dams to produce electricity. But at the same time, water is always changing radically in form. The ways in which water runs in society and is socialised without changing its character make the nature–culture dichotomy, and the way it has been portrayed and delineated, both unclear and not applicable. By virtue of its very existence in nature and society, water refutes the manner in which the dominant dichotomous distinction between society and nature has been drawn, yet at the same time it makes it fruitful to operate with another distinction: that between a natural layer and a layer influenced by human modification, or a waterscape influenced by both natural and social variables.

This opens up what can be called a hydro-historical approach: a cross-disciplinary method utilising all kinds of data – from traditional archaeological and climate data, GIS watershed modelling used in reconstructing past water-society relations, to palaeontological, hydro-logical and geological data, making it possible to reconstruct the long history of river basins, underground aquifers, precipitation and evaporation patterns, as well as different types of written sources, and so on. In practical research these enormous and complicated systems must be spatially delineated, decided and defined by those aspects conceived as relevant to social development, and can thus form bases for comparisons in time and space.

Second layer
The second layer of the analytical approach here called the water-system approach captures and highlights the anthropogenic changes in the way water flows through the landscape. Water control and water utilisation are a major aspect of most societies. They form a very wide area of activity, ranging from the human impact on the hydrological cycle, evaporation patterns and forms of precipitation, river modification schemes and the digging of canals and the construction of dams across valleys, to the millions upon millions of pipes beneath cities for drinking and sanitation, and the carrying of water in jars that so evocatively represents one of the first signs of settled agriculture. It covers everything that humans have done, and do, to bring natural water to and from their settlements – in all sectors and for all purposes, including protective measures to prevent water from destroying or undermining communities, technology, transport routes, and so on. This layer enables us to make systematic comparisons of river and water modification projects, small- and large-scale irrigation and drainage projects, sewage and canal systems, run-off

regulations, the organisation of river basins involving different countries, regions, places and cities, water consumption patterns, etc. – in both time and space. In the modern world, human modification of water systems is particularly striking, even though in many places the water's lack of naturalness is masked by the way in which the river has been engineered – beguiling because it seems so natural, but made possible because water by its appearance does not signal or reveal to where it belongs.

Water and society are now deeply interwoven, and many natural processes in the water cycle are influenced by humans; but even so, there are still river basins (both large and small) that have not been subject to human intervention, and there are enormous underground aquifers, underground river systems, cloud systems and precipitation patterns that remain unaffected by humans. The hydrological cycle does not reign unimpeded any more but crucial elements of it have evaded human control or interference, and it is this 'struggle' between the natural and the cultural, becoming an ever more important aspect of the relationship between water and society, that this two-layered approach can make intelligible in a systematic and unbiased way.

By integrating description and analysis of the two layers, it becomes possible to produce a narrative that acknowledges how many existing waterscapes are the product of both long-term and short-term cumulative interactions between human purpose and hydrological and other natural hydroprocesses. The water-system approach makes it possible to analyse the relative importance of the two layers, and how they are related. Both the layers and their interactions have effects on limits and patterns of action and their combined product will reflect the natural waterscape and the economy and technological level of society. A framework that encompasses these two layers and their relationships makes the analytical approach neither nature-centric nor anthropocentric but rather enables this crude dichotomy to be avoided in practical research.

A focus on these two layers and the relations between them will be able to capture how diverse physical water landscapes have supported the location of societies in the first place, and produced and reproduced different potentials for, and limitations of, development and simultaneously enabling analyses of how the same, particular water environment has been 'appropriated' and controlled by these same societies for the sake of particular demands and reasons at different junctures of its development. The benefit of analysing systematically both these layers is that it becomes possible to factor in how most societies at any specific point in time are enveloped by both an engineered waterscape and a waterscape that mirrors, to various degrees, the local character of the hydrological cycle. This approach also enables comparative analysis of how societies on the one hand have always had a need for water for various purposes in one form or another that their particular waterscape is expected to fulfill, and that due to population growth, shifting economic and social activities and technological capabilities the trend will tend to put greater and more multi-faceted stress on water resources. It will thus make it possible to capture how the growing multi-functionality of water, both as a physical resource and a social good, is a central aspect of long-term human history.

A specific and systematic focus on modifications of waterscapes will take into full account the economic, cultural and political importance of the diversifying roles of such actions. Water has always been an unevenly distributed means of maintaining and creating hierarchies and has thus functioned as a structuring principle in society. In some societies, control of water has been at the very heart of state-building processes and imperial legitimacy since time immemorial. Dams and large hydraulic systems are not mere technological installations: they are symbols of power. In many cases the conquest of water has served as a potent example of how some people have been able to use power over nature as a means to subjugate others. Huge water control installations clearly have economic, cultural and political importance, and their centrality and scale reflects their national standing. In some areas of the world – particularly in the dry Middle East, where water control has been especially important throughout history – dams have often been named after state leaders because few things there have as much potential to bestow prestige and authority. Similarly, since time immemorial, fountains have been symbols of urban life, distinguishing the city from the natural hazards that dominate rural life. Fountains, usually placed at the very heart of the city, have had many functions, but one of them has surely been to symbolise humanity's control of nature – a manifestation of societies' appropriation of the forces of nature; the unruly element tamed to serve the human need for aesthetic beauty.

The analytical purpose of these distinctive though interconnected layers can be made clearer by contrasting it with how the more commonly used term 'built environment' is understood. The 'built environment' is normally regarded as a product of the culture of a society, and is therefore analysed as applying solely to the socially constructed environment. The modified waterscape should on the other hand be seen as a reflection of 'culture' but also as a product of the physical character of the waterscape. The actual water that flows in a 'built' river or through a canal must therefore also be analysed in terms of the physical water context of its location, and this location's particular tradition is the product of local hydrology and geology, past water control measures and entrepreneurial action, factors which in turn, of course, are located within broader natural and societal relationships and rhythms. The relations and distinctions between the physical waterscape and modified waterscapes should also be understood as something very different from, and more complex than, the widely used pair of concepts 'managed' resources' and 'not managed' resources. It is impossible to define clearly what constitutes 'managed' resources and 'not managed' resources, because their meaning will vary from time to time and from place to place, and cannot therefore be used as a basis for comparisons or precise analyses. Moreover, the term 'management' carries a modern connotation and is somewhat out of time and place if the subject of research is, for example, adaptation and modification of water landscapes and local hydrological cycles at the time of the hunters and gatherers. People might also disagree on whether or not a particular controlled water body is 'managed'. On the other hand, provided that the necessary data are available, it is possible to reach agreement about whether a body of water has been modified or not, although there will always be disagreement regarding the degree to which it has been altered and whether or not the results have been beneficial. The two-layered approach takes as a starting point the fact that water is the same both in nature and in the most modern cities; it is the same substance that runs through distant forests as out of the tap. Water as observed in societies is both material structure and a cultural product, thereby underlining the fact that definitions and concepts of materiality in general must not be reduced to mere matter or to a static 'foundational' structure; water is always in flux and forms part of a dynamic social process. This contradicts directly the conventional viewpoint that argues that matter is a part of the natural world and thus only acts upon itself, whereas man is a human, self-conscious subject that acts upon nature and society.

The physical and man-made layers of 'open and complex water systems' underline the need, and provide a framework, for analysing how the flow of natural and social water through social space has played a pivotal role (even if occasionally in opposition to each other); one and the same water resource may have acted as a blind force of destruction via flooding, and as an encouragement to the organisation and mobilisation of co-operation and urban technological development. Water has both caused disease, squalor and human misery, and provided the means to battle these very same problems. From the familiar space of the bathroom to the buried space of the sewer, from the sparkly drops in a fountain to the tamed but still powerful force contained by dams and reservoirs, water provides a link between material and immaterial aspects and dynamics of social development. The approach can also capture how this human-modified waterscape in its turn changes the physical waterscape in an everlasting cycle of mutual interaction. It acknowledges the fact that most waterscapes are not completely natural and no waterscape completely controlled. Water expresses a paradox in nature–society relations: development presupposes modification of the natural waterscape and water always escapes its

11

developers as it evaporates back into the hydrological cycle. The same paradox gives a particular context of analysis of the long term: the most sophisticated hydraulic structures are the most vulnerable to dramatic changes in the climate or in the hydrological cycle.

By giving due weight to anthropogenic initiatives in changing the waterscape, this analytical approach to nature–society relationships appreciates the roles of the 'entrepreneur' and of human action. Interventions in and efforts to control waterscapes can in particularly dramatic ways change fundamental social as well as physical structures, both in the short and long term. Historically, individual water engineers, planners and 'water lords' have radically changed the nature of physical water systems, be it rivers, waterfalls or lakes and, by so doing, they have also changed fundamental societal structures and institutions. Water control structures can revolutionise the way water runs both in nature and society, and can thus transform societies in their very core and also diversify social developments in entirely new ways, as exemplified by the aqueducts of Rome, the Canal du Midi (which linked the Mediterranean to the Atlantic via the River Garonne in France), the Grand Canal in China, the High Aswan Dam in Egypt (which created new cultivation seasons and electrified Egypt) and the Panama Canal (which crosses the Isthmus of Panama and raises ships up to the artificial Gatun Lake). Much social science has become an abstract science of general spatial relationships, often without reference either to nature or to a subject. By including the two layers as part of the same analytical exercise, it becomes possible to analytically incorporate the creative power of human actions and aims, while still look for deeper structures that influence and constitute societies and their patterns of development. By integrating these two layers in the analytical process, this approach enables us to focus on structures while avoiding the writing of a history without subjects, or describing a society without actors and their intentions, or a nature without humans.

Focusing on the relationship between these two layers also enables us to examine and better understand a paradoxical historical trend of great and yet unknown consequences. On the one hand, more and more river systems and water bodies are the product of engineered interactions between physical water sources and human agency, but, on the other hand, societies are simultaneously becoming ever more vulnerable to substantive changes in the way water runs in nature and society.[6]

Third layer
The third layer of the water-system approach recognises and focuses on how water as an element of nature and society – as a natural resource and a social good – will always be culturally constructed and filtered.

It is concerned with how water is ascribed different meanings and has symbolised different things, from time to time and place to place for different actors (see Tvedt and Oestigaard 2006 and 2010). The history of the ideas of water has not yet been written, and what this approach underlines is that it is important to understand how these notions reflect and impact on both the physical and modified layers of water-systems, but that they also should be seen as something much more wide-ranging than those expressed in actual water control technology or water architecture.

It is crucial to acknowledge that societies' and people's ideas of water have been developed and formed in relation to a broad range of issues, water as a means of exerting social and cultural power, as an object of management practices, as religious and cultural symbols or objects, and as a signifier of social and cultural distinctions. Water has, moreover, always been used as a metaphor, most likely in all societies, although in various ways. It has been widely used as a metaphor for the stream of history and as the end of all things; it may stand for both youth and age, for power and timidity, for the female and the male, for strength and tenderness. The variations and contradictions of metaphors reflect the fact that humans' relationships with water differ both in space and time and that water plays central though different roles in people's lives.

The special character of water makes it a unique medium for cultural constructions and metaphorical traditions. Since water is at the same time particular and universal, nature and culture, physical and ideological, uniting and separating, giving life and taking life, it has been a phenomenon to which people naturally ascribed meanings. The holy water for rituals such as baptism, ablution or purification belongs to a different world of meaning from the water involved in a river's annual inundation for irrigation, or the water that nomads draw from wells in the oases, or the snow used to build igloos, or the water stored in dams for hydroelecetric power generation. But from nature's point of view it is the same water. To what extent are these cultural manifestations and elaborations of the same H_2O the result of cultural diffusion or the outcomes of interactions with different types of water? The ways in which the water worlds or waterscapes are used practically, interpreted symbolically and ascribed values according to local and regional traditions and norms have to be analysed as a result of the continuous and long-term anthropogenic interaction and mediation of cultural and natural variables in the society-water systems.

Peoples' ideas about water and how water is crucial for identities and values in a broader cultural context should be analysed in relation to which types of waters are present, or in which combinations they occur at a given time, because the different waters and their constellations are actively incorporated into the collective body of knowledge, in turn because water matters for humans at many levels (personal, societal and religious). More meticulous cultural analyses at micro-level in combination with ecological variables open up a vast ocean of hypotheses about how water has structured values, norms and hierarchies. The ever-changing qualities, capacities and forms of water enable it to function as a medium whereby we can express and negotiate social relations and problems, and communicate about the world we live in to ourselves and others.

A study of the history of conceptualisations of water must also be a study of water in religious thinking and rituals. This book suggests that a study of the role of water in religion and myth amounts to a comparative history of religions, since water plays such an important part in most people's ideas about divinities. Water is part and parcel of the history of the cosmos in most religions and provides an almost universal arena or medium for religious practices. The water-system approach argues for the need to break out of the conventional analytical framework of nation-states and civilisations in analysing ideas and cultural constructions. The reason why a focus on the ideas of water must depart from this tradition is partly that diverse water-society relations and water-society systems do not necessarily coincide with state-borders or cultural boundaries. Additionally, many notions about water are shared by a number of religions and geographical and climatic regions, so specific civilisational or cultural frames of reference are not particularly helpful in this regard. The idea that God punished humankind with floods, for instance, is shared by Judaism, Christianity, Islam and many traditional religions (Allen 1963; Leach 1969; Dundes 1988; Kramer and Maier 1989; Cohn 1996). In order to explain the complex relationship between the structuring role of particular and different human/water situations on social constructions of water on the one hand, and diffusion and acculturation regarding ideas about water on the other, a comparative and historical perspective is needed.

By operating with this distinct layer dealing with ideas about water, research can also acknowledge that the differences in how water is understood are one of the most conflictive issues in the contemporary world. In transnational river basins ideas about how the shared body of water should be harnessed are crucial to understanding regional politics and power plays between different stakeholders and upstream/downstream users and states. The strong alliance between water engineering bureaucracies and modernising politicians and their instrumental view of water has obviously played an important role in many countries in the last 150 years, and constitutes an important aspect of the history of ideas and of modernisation in general. The worldwide political schism with regard to big dams reflects different ideas about water and what it should be used for, as well as conflicting opinions about the role of water in society. In recent decades, the ideas that water should be seen as a normal market commodity and as a universal human right have provoked unrest from Sri Lanka to Africa. Finally, the global movement for 'greening the rivers' and protecting wetlands has also advanced important ideas on water that have had a great influence on societies.

By giving emphasis to ideas of water as something distinct from, but at the same time connected to, the physical character of water and its modifications through time, an analytical framework is provided that enables us to analyse both the differences and the connections between specific physical waterscapes (which will always be filtered through a wide range of cultural lenses), the modified and controlled water resources that exist at any given time (which will always reflect past actors' ideas about their water and how it should be handled), and religious ideas, cultural conceptions and managerial plans regarding water.

The water-system approach aims to break away from the reductionism of the social sciences and to counter those tendencies within social sciences that shrink the natural world to an empty stage on which only human actions matter, where societal development is conceived as something that can only be explained in terms of social facts. The approach recovers water – and thus nature – as an autonomous actor, and encourages research on the physical aspects of the relationship between water and society as well as urging an understanding of water as seen through cultural lenses. Such studies of the water-society cycle will be based on the notion that water as nature not only exists but changes, both of its own accord and as a result of human actions and in its many interactions with society, and in so doing not only changes the context in which human histories unfold but becomes part of human history itself. The water-system approach deals with analyses of the inter-relationships between three distinct but comparable factors in all societies on a continuum and over time. The historical trend is clear: more and more river systems and waterscapes are the engineered results of *interactions* between water and agency, but at the same time societies become more and more vulnerable to physical changes in their sources of water. While waterscapes in modern societies have usually been modified (there are still exceptions), even the most tamed river is still vulnerable to changes in nature because it is still connected to the hydrological cycle at different local, regional, global and atmospheric scales, which is the fundamental reason why, most likely, never before has so much money been spent on defending societies from the vagaries of their water sources as today.[7]

Documenting and analysing these clearly distinguishable but interconnected layers will make it possible to conduct rigorous comparative studies within an analytical framework that at the same time is adaptable and not rigid. Research should be thought of as a tripartite exercise, studying the distinct layers and specifying the interactions between water, technology and ideas, and structure and agent. No single discipline can manage this alone, and that is why a water-system approach will need input from all kinds of natural science disciplines as well as from the humanities and social sciences. The material foundation of human interactions with the waterscape or nature is given credit, without compromising reflexive accounts of human action and consciousness, because the approach fully recognises the importance of agency. Nature and environment are comprehended as material structures existing independent of human conceptions of them, but this perspective also acknowledges that nature and water are socially modified and constructed. Within this approach the natural exists but is not always or only natural, and the social exists but is not always or only social. Employing an analytical framework that covers all these different aspects of the social/water nexus might enable us to perform analyses that do not fall into the trap of mechanical determinism or voluntarism. And most importantly, the reductionist tradition, be it natural or social, can be overcome.[8]

It should be underlined that this water-system approach is very different from the quite influential socio-ecological system concept. While the ideas about the three-layered water system suggest a methodology for empirical research, the socio-ecological system concept is a system-theoretical concept where the system consists of what is described as a bio-geo-physical unit and social actors and institutions related to it. It is seen as a complex and adaptive systemic whole delimited by spatial or functional boundaries. Socio-ecological theory draws heavily on complexity and system theories and on a range of discipline-specific theories, such as microeconomic theory and optimal foraging theory, and incorporates ideas from theories relating to the study of resilience, robustness, sustainability and vulnerability (for example, Levin 1999; Berkes 1989; Gunderson and Holling 2002; Norberg and Cumming 2008; Mouri 2014; Bousquet et al. 2015). The socio-ecological system is therefore defined as a coherent system of biophysical and social factors that regularly interact in a resilient, sustained manner.

A conceptual, theoretical and empirical challenge with many notions of ecological systems is that such systems are perceived as a totality, as closed units. Changes within such systems are often understood in terms of different degrees of 'equilibriums' where radical changes may threaten the whole existence of the systems leading to their collapse. The problem is that what constitutes a system and its 'sustainable equilibriums' is a construction, or what would belong to level three in a water systems perspective, but water in nature is not restricted to closed ecological systems, even if these were 'original' or previously unchanged by humans. While the damming or draining of wetlands may significantly alter and even destroy habitats, they also create new water systems (a combination of levels one and two) impacting on humans and being impacted on by humans. A water-system approach represents a much more open-ended attitude, being not based on a preconceived and valued frame for understanding human–nature relations and what they should be, for better or worse.

The water-system approach aims to liberate research from any such presupposed or implied specific or fixed ideas about resilience, sustainability and regular interaction. The suggestion of studying water-society relations according to three interconnected layers aims at helping empirical analyses in time and space, and is not a theoretical concept presupposing certain ideas about specific systemic properties in a system theoretical sense. It integrates a focus on water as a physical phenomenon in social analysis, but is at the same time able also to handle analytically all those cases where water is not resilient, sustainable or part of a 'regular interaction'. The water-system approach is not based on general ideas about water as always being a critical resource whose flow and use is regulated by a combination of ecological and social systems as the socio-ecological concept presupposes, because water does not have to be a critical resource and can be regulated by either nature or society or by both together. The term 'water-society relations' assumes that this relationship is perpetually dynamic and complex, but again, to employ the approach does not presuppose a notion of continuous adaptation as the socio-ecological system concept does. The socio-ecological concept holds, moreover, that social and ecological systems are linked through feedback mechanisms. The water-system approach, however, is not based on any general assumptions of this nature since history is full of examples where a focus on feedback mechanisms will downplay the often revolutionary role of individual entrepreneurs in changing water-society relations or how sudden and fundamental alterations in the waterscape are often unrelated to the social. The water-system approach encourages all kinds of research in a pragmatic, open manner, while the socio-ecological concept is a theoretical model, based on a specific understanding of the relationship between ecology in general and society in general. There is still a great need for the connection of analyses of the social and the natural and their interconnections, and the contribution here is to suggest an open, non-dogmatic framework that can capture both long-term continuities and dramatic changes.

TESTING THE 'WATER-SYSTEM APPROACH'

This book is based on the idea that it will be rewarding for the social sciences to reconstruct, describe and understand water's movement and role in nature and in society.[9] The argument is that the relationship of societies with water makes for a general structure of social continuity through time, and that the triple-layered water-system concept evades the problems created both by natural or biological determinism and radical constructionism. The approach distances itself from extreme anthropocentrism in a double sense, while recognising the revolutionary role human modifications of water systems often have. The water-system approach and its concepts must, however, be tested empirically. The following chapters do that as part of what should be an unending series of dialectical confrontations between explanatory efforts and the hard, pitiless facts of history and social life.

The first chapter in this part presents a new explanation of one of the most important and thoroughly researched questions of all: why did the Western World and Britain succeed in transforming their societies to initiate the Industrial Revolution, where leading agricultural civilisations like China and India failed? The second chapter deals with European imperialism and the partition of Africa, and suggests a new interpretation of why Britain marched up the Nile basin at the end of the nineteenth and beginning of the twentieth centuries. The third chapter reconsiders urban studies and offers interpretations of the history and development of the city as a global phenomenon. The fourth chapter discusses the study and understanding of religion from a water perspective, and demonstrates how the importance of water and the workings of the hydrological cycle can be employed to analyse core religious cosmologies and myths and the diversity of religious practices. The fifth chapter revisits the whole debate about state sovereignty and questions both the 'myth of Westphalia' and ideas about the 'death of Westphalia' based on an analysis of the empirical role played by European continental rivers and the theoretical problems raised by a resource that cannot be controlled by territorial owners. The sixth chapter discusses international resource law and argues in favour of furthering both a historical and a physical understanding of the resource in question, using the Nile and a detailed study of the Nile Waters Agreement of 1929 as an example. The seventh chapter deals with climate history and climate change and argues that water is fundamental to any understanding of climatic processes in themselves and the challenges they pose for societies. The eighth chapter presents a case study of how the water-society approach can be useful in comparative and general studies of the history and development of countries.

2

WATER-SOCIETY SYSTEMS AND THE SUCCESS OF THE WEST

G lobal history has long centred on the comparative economic successes and failures of different parts of the world, most often European versus Asian regions. There is general agreement that the balance changed definitely during the later eighteenth century, when a transformation began that revolutionised the power relations of the world and brought the dominance of agrarian civilisation to an end. However, there is still widespread debate as to why Europe and Britain were the first to industrialise, rather than Asia. This chapter puts forward an explanation that will shed new light on their success, by showing that analysis of the relationship between societies and water is a crucial piece that is missing from existing historical accounts of the Industrial Revolution. It argues that this great transformation was not only about modernising elites, institutions, investment capital, technological innovation, exploitation and unequal trade relations, but that a balanced, inclusive explanation also needs to consider similarities and differences in how countries and regions related to their particular waterscapes, and how they were able to exploit and change their waters for transport and to produce power for their machinery. Highlighting regional similarities and differences in complex and multifunctional water-society systems may enable us to solve some of the empirical and theoretical problems with dominant modes of explanation as to the the long-term development that ended in the Industrial Revolution.

The extensive literature on why the West 'triumphed', with Britain in the vanguard, can be categorised into two main competing 'schools' or contrasting explanatory models: 'the European political-cultural exceptionalism school' and the 'European exploitation of "the other" school' – and what is crucial in our perspective is that none of them was interested in geographical conditions or water issues.

By the middle of the nineteenth century, it was already common-place to ascribe Europe's development in the late eighteenth century to its free, dynamic and progressive political and cultural character especially Britain. Simultaneously, Asia's lagging behind was seen as a result of its political backwardness, despotism and therefore cultural and economic stagnation. Later, an explicit religious-cultural theory became very influential, arguing that the divide could be explained by regional or continental differences in ideas and ethics. India and China did not have the cultural values that were a prerequisite for capitalism to develop while Europe had the 'entrepreneurial hero', that is, the successful businessman adhering to the Protestant worldview and ethic.[1] Others have argued that because the Islamic and Chinese civilisations did not develop modern science, they were unable to come up with the key technologies that underlay the Industrial Revolution.[2] Yet others emphasise how differences in law and political economy, capital and finance, education and knowledge, institutions, bureaucratic ideas, practices and markets,[3] would lead to more efficient resource use and provide for greater incentives to make investments that in turn would raise *per capita* income.[4] These explanations all reflect fundamental examples of what can be called European political-cultural exceptionalism. The essence of this school has been summarised in the one-liner: 'Culture makes all the difference'.[5]

The other dominant explanation rests fundamentally on the opposite argument: that Europe and Britain had no special cultural or institutional advantages. This school of thought interprets their success in the light of what might be called 'the European exploitation of the other' theory, arguing that it was a combination of unequal foreign trade, colonialism and the slave trade that hoisted Britain and helped fuel industrial investment and development in Europe.[6] The view that emerges from this literature is that before the Industrial Revolution, other parts of the world, particularly regions of China and India, were quite well developed and experienced similar patterns of Smithian growth to those of Europe.[7] In short, there were no internal social, political or cultural differences between Eurasia and Europe that were crucial enough to explain the divergence in development that eventually took place.[8] Thus with all other variables equal, Europe's triumph can be explained by its privileged access to overseas resources and the fact that the Europeans benefited from these fruits of overseas coercion. There exists an extensive literature in support of the general thesis that the West had no internal advantages of note, but highlights the existence of different coercive or exploitive mechanisms which account for Europe's and Britain's leadership.[9]

It is necessary to investigate these two models of explanation further to increase our understanding of the importance and ontogeny of internal political-ideological developments and external relations, but the aim of this chapter is to argue that it will be more fruitful to look up the blind spots these perspectives have in common. The most influential explanations of the 'Great Divergence' between Europe and Asia share essential structural and reductionist problems.[10]

First, by underlining either only European uniqueness or European exploitation of 'the other' they can hypothetically shed light on why Europe succeeded and Asia failed, but not why Britain assumed the leading role within Europe. Recent research has added that differences in economic thinking, 'entrepreneurial spirit' and levels of scientific development were not systematic and substantial enough to be able to explain why Britain or parts of that country 'took off' instead of France or the Netherlands in the eighteenth century.[11] Secondly, the conclusion that there were no important differences in terms of the degree of development between Britain and China or India, or north-western Europe and Eurasia cannot be taken as evidence for a conclusion that there were no important differences of relevance for development between these areas.[12] Thirdly, the debate over the 'Rise of the West' should not be a debate only over which factor or factors in British history led Britain to diverge from pre-industrial civilisations, but should include a debate over which factor or factors made it difficult for India, China, the Ottoman Empire, Russia, France, Italy and Germany to do the same at the same time; that is, it must be able to explain both transformation and the lack of it. Thirdly, a plausible interpretation should not turn correlations into causal explanations, but rather distinguish between necessary and sufficient causes, causes and occasions, and causes and prerequisites. Such an approach should also be exogenous, in the sense that fundamental technological and economic transformations need a causal element that does not itself require an economic or technological explanation. And fourthly, a convincing explanation should also make sense of what in reality was the Industrial Revolution's gradual and regional character. Moreover, a useful account of 'the triumph of the West' must break away from an almost exclusive concern with the West, to the neglect of Asian economies and technologies. Recent findings question generally accepted assumptions about fundamental differences in economic and technological levels between Europe and 'the rest', as well as between the modern and the 'traditional' West.

COMPARATIVE STUDIES OF INTERACTIONS BETWEEN SOCIETY
AND WATER AND THE INDUSTRIAL REVOLUTION

The emphasis of this book on the fruitfulness of comparative analyses of complex, multifunctional water systems in order to understand the regional and global transformations of the late eighteenth and early nineteenth centuries is not an attempt to manufacture a single-unicausal, deterministic explanation of the Industrial Revolution. This perspective does not imply a rejection of analyses that deal with overarching trends in economic relationships between Europe and Asia, or of cultural and ideological traditions, but provides rather a new context in which also such issues can be studied comparatively. At the same time, it emphasises variables that have tended to be neglected in the literature. Without in any way claiming that the Industrial Revolution was programmed to happen when and where it did, this chapter argues that it is necessary to understand how different waterscapes and water-society systems created different possibilities for the development of trade and industries in the centuries and decades before the Industrial Revolution broke through. The emergence of a group of British industrial entrepreneurs and canal builders that managed to transform Britain was not historically inevitable, but we argue that, whatever the strengths of the market orientation, capitalist mentality, or investment capital, similar entrepreneurs could not succeed in the core economic regions of China, India or of other European countries of the time, because of the nature of their waterscapes and the established water-society relations.

Transport and water-society systems
Transport has been widely recognised as a crucial factor in the Industrial Revolution, and there is little doubt that waterborne transport was of decisive importance for the new industries. Transport systems encouraged commercial expansion, facilitated the division of labour and linked production to markets. Until the middle of the nineteenth century, rivers and canals were essential in determining which regions and cities could trade with each other, and where industry could profitably be located, coal mined and iron refined. Most importantly, they allowed the shipment of heavy raw materials like iron and coal from their points of extraction to industrial sites.

However, comparative studies of the Industrial Revolution have paid insufficient attention to waterborne transport, even though Adam Smith, in his *The Wealth of Nations* (1776), recognised that industrial development depended on the infrastructure of water transport. Some studies have shown the importance of waterborne transport for the development of

Britain (Phillips 1803; Aldcroft and Freeman 1983; Szostak 1991; Turnbull 1987). Deane (1979: 76) put it thus: 'If Britain had had to depend on her roads to carry her heavy goods traffic the effective impact of the industrial revolution might well have been delayed until the railway age.' What this chapter will do is to compare the transport system of Britain with other countries and regions with a focus on the second half of the eighteenth century.

The importance of waterborne transport in the initial, decisive phases of the Industrial Revolution is underscored by the fact that most goods in pre-modern economies travelled by land, or on boats in coastal transport (Braudel 1990). Road transport was generally preferred when it came to transport of people and light and expensive goods, because it was more reliable than inland water transport, in spite of roads often becoming impassable due to rain. Precipitation could be a nuisance all year round in parts of England, while in China and India the monsoon rains made large parts of the road network unusable, even for small-scale commerce and transport of light goods, on a regular basis and for months on end. But even where and when roads were passable, available modes of road transport could not satisfy the emerging demands of the new industries. Only boats and barges could handle heavy raw materials like coal and metal ores in large quantities, as well as lime, sand, manure, general merchandise and agricultural produce and transport them at an acceptable cost to emerging regional and global markets. The development of the coal and iron industries and the success of the cotton industry depended upon improved transport infrastructure. Goods, whose price had been unable to justify the cost of transport, were now moved night and day on new water routes. Thus, comparative questions ought to focus on which countries or regions had a suitable physical waterscape where artificial waterways could most easily be built to serve such purposes, and entrepreneurs, engineers and politicians with the experience, competence and will to create such a transport network?

To the extent that transport systems have been compared in the literature on the Industrial Revolution and the 'great divide', the argument has been that if anybody had an edge it was the Chinese. It has been suggested that it 'seems very hard to find evidence of a European advantage in transportation' (Pomeranz 2000: 35). On the contrary, the 'remarkable development of water transport' in China has been emphasised: East Asia had an 'overall advantage in transport'.[13] Or again, China had a 'superb system of waterways' (Pomeranz 2000: 185). As Fernand Braudel wrote, quoting Father de Magaillans, 'No country in the world [...] can equal China in navigation' (Braudel 1979, I: 421). These assessments are wrong.

Comparisons about transport development should not treat navigability as a simple binary concept, based on the extensiveness of a canal system and the number of boats on waterways at a given time of year. When the potential and role as trade routes of the rivers and canals of different regions are compared with regard to the specific transport needs that emerged in industrialising economic sectors in the second half of the eighteenth century, many issues related to the first layer of the water-system approach need to be analysed. These include the velocity, height and frequency of rapids and cataracts of rivers, their maximum and minimum flow, tidal versus non-tidal characteristics, peak current speed, annual and seasonal variations in water level, and silt and sedimentation load. Such factors decided the cost of transport, how many times goods would have to be loaded and unloaded, the regularity of the transport system, and the extent to which goods could be shipped all year round, up- and down-stream. Furthermore, the waterscape determined the types of boats that could be used, and hence the weight and amount of goods that it was possible to transport. Other issues of significance were the relative proximity of navigable rivers and canals to new sources of raw materials, and the character of the interfaces between the sea and navigable rivers. It would also be difficult to overestimate the importance of good and stable harbours close to a navigable river mouth for the cost and efficiency of loading and unloading sea-going vessels.

The character of the second layer of the water system – river modification and canal building – was more dependent on local physical and hydrological conditions than has been acknowledged in historical writings and in the travel accounts on which they often rely. A functioning canal needed an adequate supply of water for the summit level, and open canals (all of them at that time) depended on evaporation rates and on soil types that did not leak. Ensuring a sufficient all-year water supply for the canals was the main problem in most areas, as this depended on the seasonal variability of rainfall, glacial melting and the layout of river systems. Floods also washed away embankments, covered locks and made access difficult. The work and investment needed to construct and maintain artificial waterways were, however, not only influenced by waterscapes, but also by political relationships among the various users of rivers and dominant ideas about water.

These factors combined would decide the extent to which inland water transport was possible or sensible, or indeed economically viable at all. No systematic comparison of waterscapes, hydrological cycles and transport potentials in the major regions of Europe and Eurasia has been carried out in relation to the 'Great Divergence' debate until now. This chapter presents some data from such a research project, but focusing mainly on Britain, China and India.

WATERSCAPES, WATER-SOCIETY RELATIONS
AND TRANSPORT IN BRITAIN

Compared to the core economic regions of China and India, Northern Italy and France, parts of England and Scotland possessed a unique system of relatively easily navigable waterways, and waterborne transport had long been more reliable than road transport. Moreover, medieval water transport was cheap, for carriage by land could cost ten times as much (Jones 2000). The rivers were fed by rain throughout the year, and there was relatively little variation in water levels. The rains that created muddy, impassable roads made rivers navigable. The rivers had few rapids, and the waters carried little sediment, and they did not normally freeze in winter. On the coasts facing the North Sea and the Irish Sea there were estuaries with tidal rivers that penetrated far inland, creating sea routes and land routes simultaneously, long before the advent of steamers being capable of moving upriver against the current. The high number of such rather short and narrow rivers became the commercial lifelines of the medieval kingdoms, encouraging early regional specialisation and urbanisation. British and especially English rivers were remarkably reliable, especially compared to those of India and China, even though Britain could also experience floods, droughts and freezing rivers that caused problems for the boatmen and for trade in general. Compared not only to India and China, but also to France, in England it was easier to dredge rivers, reinforce banks, straighten and shorten river courses and control water levels with sluices and staunches of modest size and complexity. Britain's river system and precipitation patterns were so benign that waterborne transport could largely rely on natural water routes until the 1750s.

England also had a crucial advantage when it came to perennial water supply for small artificial waterways for two reasons; sea-level canals in moist areas had a high water table and most of the country had year-round rainfall. While Britain had very few proper canals as late as 1759, in less than a generation the whole face of England was furrowed with navigable waterways, and bulky goods were more and more carried by boat. It has therefore been aptly stated that 'the map of English canals, is the map of industrial England' (Hartwell 1967). Canal boats could carry 30 tons pulled by a single horse, or more than ten times the load per horse drawing a cart. New technologies, notably the use of puddled clay, and new types of locks and boat lifts, also encouraged canal building.

The 'canal mania' after 1760 until about 1830 when the railway came, gave high returns to a new class of shareholders. It has been argued that the canal building ventures even created a new kind of business mentality, or the 'true beginnings of financial capitalism', utilising surpluses from agriculture and trade, and capital provided by manufacturers and other investors (Bryer 1999: 687; Bryer 2000: 158. See also Baskin and Miranti 1997: 127; Bagwell and Lyth 2002: 12). The technological, financial and political challenges in canal building were within the reach of private entrepreneurs, while in countries like China, India and France, the state was directly involved, as the physical challenges presented by their respective waterscapes were such that they could not be solved by individual citizens or small groups of businessmen. The canal system in England was developed into an interconnected web of quite efficient waterways that could be used on a quite regular all-year basis. Very importantly, both coal and iron deposits were available within the ambit of Britain's water transport system, to a much greater extent than in any other country. This made it possible for raw material sources to be linked to the production sites as well as to the market, and every city in England, except Luton, was connected to the sea by 1800.

To summarise: England possessed 1,900 kilometres of navigable rivers in 1725 (Willan 1964: 13). By the end of the 1820s, the country had 3,400 kilometres of navigable rivers, and 3,200 kilometres of canals (Clough and Cole 1946: 446). Waterborne transport had thus become more crucial than ever, and bulk goods such as coal and iron could be transported at a much lower cost than by road (Jackman 1916; Willan 1964; Bagwell 1974).

WATER SYSTEMS AND TRANSPORT IN CHINA AND INDIA

Chinese rainfall patterns caused extreme annual and seasonal fluctuations in river discharges that were scarcely comparable with European rivers, in particular with those of Britain. Moreover, marked seasonality and long dry spells profoundly affected the availability of water for canals (Yu 2002). Extreme variations in river levels were common, as were high rates of bank erosion and sedimentation. The colossal human efforts needed to protect societies against the physical characteristics of violent rivers and monsoon climate translated into serious impediments to the development of transport infrastructure. Furthermore, run-offs from the principal river drainage basins are gathered into remarkably few main outlets to the sea in China, with only some five to six outlets for about 4,000 kilometres of main coastline to carry off enormous river discharges, fed not only by the monsoon but also by Himalayan glaciers. The characteristics of these widely separated river systems differed greatly, but they could all cause catastrophic floods, and indeed they all flooded very frequently. Commercial centres were worst affected, because they were usually located close to these rivers and on vast, fertile delta plains, partly because rivers were the main transport arteries of agricultural goods.

Large river engineering schemes, which run like a thread through Chinese history, were implemented primarily to serve agricultural production, as well as military and administrative goals. Within the same social and dynastic structure they could not easily be tailored to meet the needs of new industries and global trade. Water management traditions, or the water planners' habits of thought, had for generations been concerned first and foremost with taming water, that is, with drainage and defence against recurrent and dangerous floods and droughts. The relatively advanced multifunctionality of the Chinese river system most likely discouraged regional and local attempts to use water primarily for transport or mechanical power production. China definitely possessed the technological capacities to build canals.

The Chinese state can most aptly be described as a 'water-moving-state'. Since its very beginning the strength and legitimacy of the state depended on its ability to control and move water, and it organised great water projects that involved canal building and modifications of river systems on a grand scale. The most famous of these canals was, of course, the Grand Emperor's canal. The oldest parts of the canal date back to the fifth century BC, although the various sections were finally combined during the Sui Dynasty (581–618 AD). The total length of the Grand Canal is more than 1,700 kilometres and its greatest height is reached in the Shandong at a summit of 42 metres. It connects the Yangtze River with Beijing, passing through Tianjin and Hebei, Shandong, Jiangsu and Zhejiang provinces.

Since the most highly economically developed centres of Chinese civilisation in the seventeenth and eighteenth centuries were located in the Yangtze and Huang He basins, it is natural to compare the industrialising regions of Britain with these two river regions. The Yangtze River was China's main trade artery, and economic progress was directly connected with the intricate network of waterways (Elvin 1977). The river basin had more than 20,000 kilometres of waterways, and Marco Polo noted that thirteenth-century official reports indicated that 200,000 boats descended the river annually (Wiens 1955: 248). The many canals were one of the most characteristic features of the region. Transport by water was so much cheaper than by land that areas without water transport were usually much less developed. The difference between the two modes of transport was so pronounced that it has been seen as 'a case of premodern economic dualism' (Elvin 1973: 304).

However, the Yangtze plain was exceptional and the regional water transport network was dense. Even so, it was incapable of efficiently conveying heavy materials. Iron and coal were mostly found outside the river basin, or upstream of the Three Gorges, and were thus located in areas that could not be served efficiently by water transport. The transport network had also clear limitations regarding even regional transport of locally produced goods. First, the layout of the canal system had not originally been to serve the exchange of goods, but mostly for drainage and flood protection. Moreover, the water level in canals fluctuated with seasonal variations in rainfall and river discharges. They were thus often not designed to move goods efficiently all year round, but to move tribute grain at certain times of the year when they were navigable. There were also huge variations in river water levels, with dangerous seasonal floods alternating with low waters. Silting of rivers and harbours and meandering tributaries on the plains made these waterways difficult to use for the transport of heavy raw materials and industrial products. Boatmen and traders in eighteenth-century England complained about the irregularity of the Severn and the Avon, but, compared to Chinese rivers, they were more like predictable highways.

As for the Huang He (Yellow River), the 'mother of Chinese civilisation', flooding transformed it into a raging torrent, carrying 88 times more water than in the dry season. The river had four natural flood seasons, which made regular large-scale transport extremely difficult. A detailed reconstruction of the behaviour of the Huang He from 1650 to 1850 remains to be made, but it has been estimated that between 602 BC and 1949 AD there were more than 1,500 major floods, 25 significant channel alterations and seven major changes of course. Like other mighty Chinese rivers, the Huang He originated high in the Himalayas, subsequently crossing extensive alluvial floodplains, and a massive transport of sediment contributed to the shifting of its channel. The river mouth thus moved from north to south and back again, with the most recent 'natural' shift occurring in 1855, when the channel mouth moved from the north side of the Shandong peninsula to the south (Zhang et al. 2010).

The interface between rivers and sea was much less conducive to trade in China than in Western Europe. China's flat coastal landscapes were formed by the deposition of fine river sediments and the river mouths were often choked with silt, and time and time again their outlets to the sea would shift. There was a lack of tidal rivers, and riverbanks were frequently sandy and shifting. Access to the sea from the Yangtze was not usually a great problem, but one major channel to the sea did not constitute a dynamic inter-regional transport network, similar to the one that could be developed in England. Although maritime trade was substantial, and the Chinese had the technological capability to develop seaborne transport, the hazards of the maritime route were both natural and societal, including the problem of piracy. Beijing thus prioritised the Grand Canal and inland water routes.

Clichés about the inward-looking, conservative 'oriental mind' or 'stagnant' Chinese civilisation are highly misleading where canal building is concerned. The Chinese carried out many major projects from the third century BC onwards, developing sluice and lock technologies and constructing a myriad of canals, the most famous of which was the Grand Canal, which was more technologically sophisticated, administratively challenging and managerially complex than ever by 1800 (Dodgen 2001).[14] Use of the canal was limited, however, due to silt and water level problems, and by its narrowness, which made it unsuitable for heavy loads. Indeed, the Grand Canal complex collapsed in the nineteenth century, as it silted up from the mid-eighteenth century. Despite the experience of Qing engineers, the state did not have the resources to control the canal when changes in climate altered the amount of water carried by the Huang He (Leonard 1996). Nor was there sufficient relatively silt-free water at a suitable altitude in Shandong to keep the canal functioning throughout the year.[15]

Moreover, the navigable rivers and the existing canal system were far from where the most important coal and iron deposits were to be found. China's vast river system, characterised by river basins whose water runs more or less uni-directionally from west to east, turned rivers into barriers to north–south trade in general and to the transport of heavy goods in particular. The huge rivers were often difficult to cross by boat due to their highly variable patterns of discharge and high flow speeds and, unlike the many small rivers in Britain, they were also impossible to cross on foot or with wagons, since bridges could not be built due to their enormous width. Those features, which for hundreds of years had made the rivers quite efficient seasonal highways for the downstream transport of agricultural products grown along their banks, became a drawback when a regular, perennial and gentle flow was necessary for the shipment of heavy goods.

In India, monsoon rains were the most important negative physical factor, influencing navigability, reliability and opportunities for port establishment. The monsoon limited inland navigation to a seasonal activity in some areas (for an overview, see Jain et al. 2007). The major Indian rivers had been harnessed and exploited for generations, and they encouraged the development of a very productive agricultural civilisation, but as transport highways they suffered from some of the same hydrological problems as Chinese rivers. The Sanskrit word for the Indus River, Sindhu, means 'ocean', but in the winter season, when the melt water from the Himalayan glaciers was radically reduced, the river resembled rather a series of pools connected by narrow and shallow channels. On the Ganges, and even on parts of the Brahmaputra system, it was possible to move goods in the dry months from November to

January, but in the spring, many river branches became too shallow for even very small boats (see, for example, Forrest 1824). On the other hand, during the monsoon, the rivers flooded and became dangerous for navigation. In spite of these drawbacks, rivers played a very important role in traditional transport, because water transport was still more efficient than transport by land, which in many areas came to a standstill during the monsoon. In Bengal, for example, about 80 per cent of trade goods were carried by river.

However, while rivers were crucial as local transport routes, they were not conducive to growth in national and international trade. One author concluded that rivers in India 'are not considered fit for navigation' (Bharati 2004: x; see also Hart 1956). This problem can be illustrated in another way. One-sixth of the area of modern India is drought-prone, and one-eighth of the area is flood-prone, with floods rendering an average of 33 million people a year temporarily homeless (Verghese 1990: 8–9). Although the number of affected people was less at the beginning of the nineteenth century, floods and droughts could strike much harder, since there were insufficient dykes and storage facilities for flood water for use in the dry season.

India is probably the country with most river migration and 'disappearing' waterways. Indeed, the mythical and famous River Saraswati can be seen as a religious expression of this predicament (Jain et al. 2007: 870–913). A distinguishing feature of South Asia is changing riverbeds and cities left high and dry. This is mainly due to the fact that the Himalayan courses of these rivers are extremely torturous, while on the plains, due to heavy sediment loads and strong flows during the flood season, they meander across the floodplain, frequently shifting course (see, for example, Khan 2005). It is safe to say that there is no river in the Indo-Gangetic Plain that has not changed its course a hundred times, mostly due to factors that outwit humans' control. Human settlements from the earliest times had followed the rivers, which meant that every deserted river implied a disturbance of settlements, and the abandoning of villages, towns or even great cities. These phenomena removed the water supply to an inconvenient distance, destroying strategic advantages and established trade routes. Larger changes of course have had proportionately more serious consequences. Lines of mounds marking abandoned villages along the line of a former river are sufficiently common in the Punjab to have a name: *thes* (Wood 1924: 3). In India, the monsoon brought heavy rainfall during the monsoon season, producing natural reservoirs in depressions even far from the main rivers. They were replenished annually by the rains and some of the earliest settlements were established along these depressions as well as along the riverbanks. But in some regions, such depressions might not be replenished by rainfall for several years. This led to many settlement shifts throughout history.

During the first phase of the Industrial Revolution in Britain, India experienced a number of radical changes in its waterways. At the end of the eighteenth century, the Beas River was 'captured' by the Sutlej, a tributary of the Indus, and its old bed became a dry ravine. The Brahmaputra also changed course. Until 1787 it had flowed through Mymensing, but then it started to move west, while the Teesta in turn formed the easternmost branch of the Ganges (Michel 1967: 48; Bernstein 1960: 14–16). In times of flood large changes in the courses of the rivers were particularly likely to occur. The Damodar River had formerly joined the Hooghly about 50 kilometres upstream of Calcutta, but in the great flood of 1770 it left its old channel altogether, and joined the Hooghly about 50 kilometres below Calcutta. The consequence was a marked deterioration in the channel of the Upper Hooghly, and a silting-up of the old channel. This lack of river stability was acknowledged by the British as one of the major obstacles to transport improvements when, several decades later, they tried to develop steamboat traffic on the Ganges (Bernstein 1960).

Not only were the rivers and water availability in China and India much less conducive to the rapid development of water transport and facilities than in Britain, but large-scale canal construction was technologically impossible at the time. This was due to the huge seasonal and annual variations in potential sources of water supply for artificial waterways, and the enormous sedimentation problems that threatened to destroy canals and embankments in the course of only a few years. Furthermore, the permanent need to protect land and people from floods and droughts made it dangerous to establish transportation canals, while there were serious conflicts of interest between the irrigation and transport sectors regarding the type of water control works envisaged for various economic activities. The canals that existed in some regions of both countries were not suited to the transport of goods like coal and iron, since they had not been built with that purpose in mind. Nor were their rivers suitable as arteries for global trade, due to their strong seasonality and the sediment problem where they met the ocean. The boat-building technology available at the time made it almost impossible to build boats that could carry heavy goods on such seasonal, violent rivers. Large areas of both countries had no navigable waterways whatsoever, and those that they had did not run close enough to where coal and iron were found in sufficient quantities.

Trade, the instrument of any developing economy, had been hampered for centuries by transport, which was invariably slow, inadequate, irregular and, not least, very expensive. As Paul Valéry remarked: 'Napoleon moved no faster than Julius Caesar.' The water transport system that developed in

Britain during the late eighteenth century signalled a transport revolution. No other country in the world possessed such a highly developed network of waterways. Although transport was still slow, trade in heavy goods could be quite reliable. It could be carried out on an all-year basis, and boats on the small rivers and canals could carry iron, coal and sand from where they were found to where they were to be used. Only in Britain could an efficient and reliable transport system be established at the time, bringing the different regions together, and linking them to regional and global trade routes.

MANUFACTURING AND WATER-SOCIETY SYSTEMS

Whatever definition of the Industrial Revolution we use, all emphasise the centrality of the rise of the modern factory in England in the last third of the eighteenth century. Even if this is not in itself a sufficient explanation, the use of machinery remains the principal aspect relative to which every other issue is studied. Before the steam engine came into widespread use in the mid-nineteenth century, only two types of inanimate power were available: wind and water (Watts 2005: 53).

While windmills were not sufficiently reliable for the new production processes, water power made large mechanical workshops feasible, fostering industrial discipline among a modern working class. Water was the main power source in the initial phase of the Industrial Revolution. All-important industries, such as textiles, iron, paper, pottery and a myriad of less important ones were dependent upon water power of this type. A number of writers have shown that water power continued to be the most important source of industrial energy well into the nineteenth century (Von Tunzelmann 1978; Pounds 1973: 38; Kanefsky 1979). It is therefore essential to understand and analyse from a comparative perspective the role of water power, notably the reasons for its dominance in Britain and for its marginality in China, India and other European countries at the time. While this chapter deals with the role of water in the Industrial Revolution in general, the discussion focuses on the water-powered mechanisation of the textile industry, due to this industry's particular importance in analyses of why Britain succeeded and India and China did not.

While water remained the main and sole reliable source of inanimate power, the waterwheel was the key technological factor in all the basic and new industrial processes. Seen from a perspective of about 2,000 years, the first phase of the Industrial Revolution saw the zenith of the societal and economic importance of the waterwheel, which most new factories utilised to drive their machinery. For the more efficient vertical waterwheels that had been developed, the overshot wheel almost always required an aqueduct, often elevated, leading water to the top of the wheel where it poured into the buckets. The wheel was rotated by the weight of the water in these buckets. Aqueducts could be artificially made by constructing small dams and millraces. The waters needed to satisfy three conditions: a year-round supply, not too much silt and adequate stream power capable of turning the waterwheel 24 hours a day throughout the year. This particular relationship between economic activity and natural resources meant that the new factories had to be located where there was running and falling water.

Thus the fundamental question became where one could find sources of water that could easily be adapted to the need of the new industries, and that were located close enough to both sources of raw materials and national and global trading routes. In spite of the importance of this form of inanimate power in the first phase of the Industrial Revolution, no systematic comparison of water-society relations has been made.

Water-powered industry in Great Britain

Parts of England, as well as parts of Wales and Scotland, had waterscapes conducive to this type of industrial power. The great number of relatively small perennial rivers fed by all-year precipitation, such as the Derwent, Irwell and tributaries of the Severn, with modest currents and silt-free water, were perfect for existing waterwheel technology and for developing new technologies (Baines 1835: 55–83; Robson 1957: 1). For centuries, the prevailing conditions in various areas of England had made it an ideal workshop for practical engineering and experiments, which encouraged a wide variety of water-driven inventions. This disparate economic-technological milieu grew in size and importance, and, in the final decades of the eighteenth century, it turned out a number of new inventions, such as the water frame, Crompton's mule and Henry Cort's grooved rolling process and puddling furnace. It was relatively easy to channel sufficient water to where the waterwheel was situated, thanks to the wealth of experience of making minor modifications to river embankments and canal construction, and the existence of a decentralised, bottom-up water management system. But why did the water-powered mechanisation and technology flourish so much more richly in Britain in the period 1760–1820 than in what might be called its European birthplaces: parts of northern Italy and eastern France? To explain this, many factors must be considered, but one of the main reasons was the great discharge variations of the Alpine rivers; the force of the peak currents and the turbulent, flood-prone rivers such as the Rhône and the Po made it difficult to operate modern factories all year round at that time.

The textile industry was the leading sector in the Industrial Revolution, and the first to adopt new forms of industrial organisation. The most spectacular growth took place in the cotton industry between 1750 and 1800, by which time Britain had become the world's leading exporter of cotton textiles (Ellison 1968: 57–70, Robson 1957: 1–3). This power shift was made possible by many factors, political, cultural and imperial, but without the development of new factory technology, such as new spinning and weaving machinery in England, it would not have happened. When multiple spindles began to be mounted on frames, it became evident that human power was inadequate, and as long as water power was the only available substitute for human and animal energy at the time, the waterscape attained a new and fundamental social and economic importance. Indeed, yarn made this way was known as 'water-spun' as opposed to 'hand-spun'. Cotton could not be grown in England, for climatic and water-related reasons, but it could be processed there, thanks to the same features of the climate and water systems.

It is therefore symbolic that the location of the first modern factory in history, a cotton mill in Cromford near Derby that was established in 1771, was on the banks of the Bonsall Brook, a tributary of the River Derwent. Arkwright's 'water-frame' was installed here to be powered by large waterwheels. The mill swiftly grew to house several thousand spindles and 300 workers. A disadvantage of the Cromford site was poor communications, so Arkwright had the Cromford Canal built to transport raw materials and finished goods to and from the site. In 1780, the owner bought land for a yet bigger factory complex and a larger mill, and moved to a more powerful river, the Derwent itself. Fifteen years later some 140 similar water-powered mills were spinning cotton. By 1800, there were about 900 cotton-spinning factories, most of them in the North and the Midlands, 300 of which employed more than 50 workers. In the late eighteenth century, there were nearly 100 cotton mills within a 10-mile radius of Ashton-under-Lyne, all on the River Tame, and all powered by water. By 1816, the average number of workers employed in 42 Manchester textile factories was already as high as 300 (Redford 1960: 19, 27). More efficient technologies increased productivity and quality. Crompton's mule gradually overtook Arkwright's water-frame as the preferred machinery. But this was also set in motion by waterwheels, fitted with as many as three or four hundred spindles. In 1835, there were 109,626 power looms in the United Kingdom in the cotton industry (Aspin 2003; see also Bowden 1925; Daniell and Ayton 1814; Hills 1970). The cotton mill became one of the defining symbols of the Industrial Revolution, signifying material progress and the growth of industrial spirit and identity. Wool-spinning and weaving were mechanised during the same period, using water at crucial stages of the production process.

The mill and all that came with it thus marked the dawn of a new era. It was not until 1783 that the steam engine was first used in a factory, and then only indirectly (Arkwright's Manchester mill used a Newcomen engine to pump water to drive the machines in the dry season when the water pressure was insufficient to turn the waterwheel).

At the beginning of the nineteenth century, in brooks and streams all over England, waterwheels were set up, effectively powering not only the textile industry, but also the metallurgical industry. Indeed, steam engines could not have been built in the first place without waterwheels to drive the equipment that was needed to smelt the iron and form the cylinders and other metal parts of the steam engine. The development of the iron industry was crucial because it allowed the production of most of the machinery required by other industrial activities, whereas earlier machines had been made of wood. The key factors in the iron industry were the heating process and the hammering of smelted iron, both of which depended wholly upon water power. Without water-powered bellows it would simply not have been possible to produce temperatures in the blast furnaces sufficiently high to produce cast iron economically. This water-aided furnace technology was an essential element of the iron industry and its high output. The iron industry depended on water power to produce the new machines of the new industries, such as rolling mills, metal lathes, hydraulic hammers and so on, which also required water power. It is therefore safe to assert that the increase in productivity was dependent upon the more efficient use of water, in this industrial sector as in other emerging industrial sectors.

The success of waterwheel technology in parts of England cannot be sufficiently explained by European political-ideological exceptionalism, or by European colonialism and unfair global trade regimes. The fundamental technology itself was not a European invention, but was first developed in Asia, probably in India or China. It was introduced to England by the Roman conquerors. From its earliest days, the technology spread quickly throughout England. At the time of the Domesday Book, compiled for William I in 1086, more than 6,000 watermills were registered, all using the small brooks and streams that were full of water at the time of the corn harvest. The waterwheel was then primarily used for grain milling but was gradually adopted for other purposes. The mechanical fulling-mills that cleaned woollen textiles were especially important. Around 1200, water-powered hammers were brought into use in many small rural forges for beating hot metal into shape. The Wealden iron industry was located close to streams because these were where it was possible to accumulate the head of water needed to drive the hammers. Great improvements in watermill technology took place in the sixteenth and seventeenth centuries (Unwin, Hulme and Taylor 1924: 27). However, the development of the overshot system was crucial for the industrial breakthrough (Reynolds 1983). In technological terms, the equipment in itself was not revolutionary, and it would probably not have puzzled or surprised a millwright from medieval China or the Ottoman Empire. Many of the early water-powered cotton factories were converted silk mills (Lewis 1848: 95), which themselves were often converted medieval corn mills, an indication of the long and varied history of water-powered technologies.

By the end of the eighteenth century, entrepreneurs and businessmen in much of England had wide experience of using waterwheels for productive purposes. The role of and dependence on water can also explain the distinct regional pattern of England's transformation in the late eighteenth century, for industries, which often were clustered in combinations of textiles, iron and engineering, had to localise along streams and brooks. No such transformations took place in parts of England that lacked rivers and brooks with sufficient force to power waterwheels, or where it was not possible to build dams, reservoirs and man-made waterfalls that could drive these 'water engines'. For example, it mattered that about 250 kilometres of year-round, ice-free and stable rivers, streams and canals flowed within 15 kilometres of Manchester city centre. Some waterways, such as the Dene, Tib and Corn Brook, important in previous centuries, have now been forced underground, while four rivers still flow visibly through the city: the Irk, Irwell, Medlock and Mersey. It also mattered that the water ran down gentle hills, carrying sufficient energy to drive the waterwheels. After all, the Romans called this area Mamucium, the place of the breast-shaped hill, from which the word Manchester later derived, due to the contours of the landscape.

WATER-POWERED MECHANISATION IN THE
CORE ECONOMIC REGIONS OF CHINA

Explanations of the success of Britain based on assumptions about European technological supremacy are not very convincing when it comes to water control, for water power had already been an important source of energy in ancient Chinese civilisation. In fact, China led the West in this respect. Large rotary mills for grinding grain appeared in China in about the second century BC. The typical Chinese waterwheels were of the horizontal type, although the vertical wheel was known and was used to operate trip-hammers for hulling rice and crushing ore. The edge-runner mill appeared in China in the fifth century AD. The trip-hammer was in use in China perhaps as early as eight centuries before it was used in Europe, and China invented the first water-powered blast furnace. China had factory-like establishments, and iron works that employed 1,000 men (Elvin 1973: 307–8. See also Wagner 1984: 95–104). Some historians argue that iron output in the eighteenth century was greater than ever, possibly exceeding 200,000 tonnes per year (Eastman 1988: 137–47). The Chinese were certainly familiar with milling technology, as the earliest paper mills date from around 1570, the earliest sawmills from 1627, and mills for winding silk from cocoons from 1708 (Needham 1996: 4, 404, 394, 405).

It has been documented that the Chinese were using a water-powered, multi-spindle spinning machine in northern China by the end of the thirteenth century.[16] We do not know why such early Chinese machine-spinning disappeared, but a reasonable hypothesis is that it was related to changes in the waterscape (either natural or man-made) and the limited diffusion of them was restricted by a general shortage of suitable rivers and streams for this type of industrial activity. What we do know is that water continued to be used as a power source in various places in China where such water was available. The Chinese were well aware of the benefits of water power. In Hunan, water was lifted with stream-driven norias, and a contemporary poem praised how much easier this was than pedalling (Elvin 1998: 113–200, 136–7).

The Chinese knew the technology and definitely had the scientific and engineering capability to build reservoirs and dams, but the waterscape was in general difficult to regulate for this purpose, given the extremely variable flow in most of their rivers and the fact that the most important ones in the populous cotton industrial areas crossed extremely flat flood plains. Furthermore, such reservoirs would soon be filled up with silt. The Chinese rivers were extremely difficult to use for regular industrial production given the technology of the time. Chinese factories could therefore not get the regular power supply that could transform them into modern machinery-based factories. The rate of invention and application of iron instruments was much slower than in Western Europe (Elvin 1972: 137–72). A main reason was that there were very few exploitable water resources close to where iron was found and made, which affected negatively the heat of the smelters and thus the quality of the products.

The question of the Chinese cotton textile industry has been thoroughly discussed. The Chinese had the technological capacity, enough raw cotton and sufficient demand for cotton textiles to develop modern cotton textile factories, but no such factories were established before the latter part of the nineteenth century, after the steam engine had been introduced. China's textile industry was overwhelmingly located in the populous Lower Yangtze region, for long the core area of the Chinese economy, where large merchants established considerable control over the production process of better grades of cloth (Nishijima 1984: 17–79). The Ming cotton industry demonstrates that the idea of the stagnant

Chinese society is misplaced, since a family-based rural industry, with hundreds of thousands of people working in the weaving and spinning industry, had developed in the eighteenth century, stimulating important changes in production and marketing. Spinning and weaving were done by every family in the many small villages dotting the area around Shanghai (Lindsay and Gützlaff 1833: 188). Although the Chinese were aware of the watermill technology, and acknowledged its advantages, they still did not use it in cotton-producing regions. All scholars seem to agree that a critical weakness of the industry was the relative absence of mechanisation.

My argument is that the Chinese were unable to copy English textile machinery, even if all other cultural and political factors had been equal, because they did not have the rivers, streams and brooks that could power the new industry that the British had. The river control works in China of the past had, moreover, been engineered to solve other tasks and served water interests and waterworks that would compete with projects to tame rivers in order to harvest their industrial potential. On the extremely flat Yangtze plain, crossed by a violent, silt-laden river that drained 70–80 per cent of the country's precipitation, there were very few places that were suitable for exploiting the flow of water for driving waterwheels, especially overshot vertical wheels. The Yangtze and its main tributaries could not be used for producing power through waterwheels like the much smaller and more modest English rivers, streams and brooks. In major cotton-producing regions, the head of water was not sufficient to drive hydraulic machinery. Frequent flooding also functioned as a major disturbance influencing the dynamics of the river–land interface. The Huang He could not be used either, not only because of its irregularity and the fact that it broke through the levees almost every year, causing floods and displacements, but because of the amount of silt transported by the river which would damage or destroy the vulnerable waterwheels.

The fact that China did not modernise its textile industries has again and again been described as a 'great puzzle' (Chao 1977), but what has been shown here, by integrating water-system relations in the analyses, is that it is no longer a puzzle. Explaining the lack of suitable channels and waterfalls as a function of a lack of entrepreneurship, capital or competence, is misleading and confusing. Dominant explanations have focused only on various social variables and have thus overlooked a fundamental geographical constraint in China: the sheer lack of relevant inanimate power sources to power machinery.

WATER-POWERED MECHANISATION IN THE
CORE ECONOMIC REGIONS OF INDIA

Until the second half of the eighteenth century, India was the most important cotton textile producer in the world (Parthasarathi 2001). According to some estimates, it accounted for 25 per cent of global manufactures in 1750 (Bairoch 1982: 296). The key to this production process was the pit loom, a technology that endeavoured to mechanise work previously performed entirely by human labour. This horizontal loom was lightweight, consisting of only a few pieces of wood that could easily be dismantled, transported and reassembled. In the late eighteenth century, it was not unusual to see an individual weaver moving house, carrying on his back everything he needed to start weaving the moment he arrived in his new home. The pit loom's mobility was an advantage for weavers, since they could compete with other weavers at local level. However, it was at a technological disadvantage compared to the efficient factory system that was being established in England. This type of loom, however, was also totally dependent upon water, but for a quite different reason, and with far-reaching implications. The hole under the loom where the weaver sat and worked the pedals (hence the name 'pit loom'), created the humidity necessary for the cotton to be woven. In parts of South India, however, even hand-weaving itself came to a standstill for about a month because of the heavy rains. In some areas, high winds were a regular annual problem, breaking 'the warp yarns that were fixed in the loom'. In Kongunad, the work schedule was reflected in concentrating festivals for the left-handed caste, of which weavers were an integral part, in the months of the monsoon. During the rains, yarn preparation, which was done outdoors, could not be performed (Parthasarathi 2001: 12, 19).

India did not possess the necessary power sources to develop a modern factory system, before the steam engine liberated industries from the riverbanks, due to the difficulty of exploiting rivers in both the rainy and the dry seasons.[17] Moreover, the large rivers crossing the northern plains had very few places with a sufficient head of water. The Ganges, for example, falls only 215 metres on its 1,600-kilometre journey from Delhi to the sea. The extreme monsoon pattern in India brings 90 per cent of the annual rainfall between June and September, resulting in swollen, violent rivers. During the rest of the year, rivers are shallow and slow-flowing, and some even dry up completely. To use these rivers for year-round power production was virtually impossible at the time. Moreover, they ran through flat plains, whose only relief was flood-plain bluffs and belts of ravines and badlands formed by gully erosion along the larger streams. The steady or sudden migration of rivers and riverbeds made factory building on riverbanks a very risky business. In Bengal, where the

39

textile industry was strongest, rivers meandered through the slopeless plain of the Ganges–Brahmaputra delta. About 80 per cent of present-day Bangladesh is less than 10 metres above sea level, and in the flood season almost half the country is normally under water. The rivers in the southern parts of the sub-continent depended almost completely on the monsoon and were generally more irregular than the Himalayan rivers.

In India, therefore, inanimate power was mostly used only in a few places where watermills could be established. There were watermills along some of the streams of the Deccan plateau, where Panchakki took its name from a seventeenth-century watermill, which used to grind grain for the pilgrims and troops of the garrison. A mountain spring provided the water which powered the mill. Water power was also used in the north-western district of Hazara 'to manoeuvre a wooden trip-hammer for milling rice'. The vertical wheel was also found in a water-driven cotton gin in another part of the same district. The latter, it has been argued, was a local invention (Raychaudhuri 1982: 261–308, 292–3). Hazara, unlike the rivers on the plains and in the Deccan, had sufficient water to drive a mill, to a great extent on a perennial basis. These few pockets of water power make it clear that the technology was known, and they indicate that the problem was the diffusion of this technology to the very different water regimes that dominated in the central economic regions of India.

A NEW INTERPRETATION OF THE 'GREAT DIVERGENCE'

This chapter has shown that an inclusive explanation of why certain regions of Britain were the first to industrialise must consider and analyse the complex, multifunctional water systems within which these developments took place. The reason is that before the dominance of the steam engine, the revolution in transport systems and the development of the modern factory system that transformed much of Britain from the 1760s to the 1820s were linked to changes in how the British related to and used their particular waterscape. It has also described how the central economic and political regions of India and China did not have water systems that could be used or developed as easily and profitably as they could be in parts of Europe. Variations in hydraulic designs can thus not only be interpreted in social terms or reduced to social variables.

Water-society systems affected other important variables relevant to growth and development that have not been specifically dealt with here, for example management practices and ideas about water. In the latter half of the eighteenth century, Chinese water management traditions could be categorised into four main activities: drainage, flood control, irrigation and transport. Drainage had been an important activity on the flood plains, due to the combined impact of the monsoon, melting ice in the Himalayas, and the flatness of the land. Even Chinese creation myths were connected to mythical stories about drainage; the Emperor Yu became mythical because he managed to drain the lowlands in the mythical past. Flood control has been a key task throughout Chinese history, particularly on the lower reaches of the major rivers. In some cases, therefore, navigation canals were a sort of fringe benefit of proper levee maintenance. The gradual development of a system of drainage canals, and the water management system that was established to oversee their maintenance, was more a hindrance than a benefit to building canals for transport of goods and for turning the rivers into power sources for machinery. The density of irrigation canals in both the Yellow River and the Yangtze basins also slowed the development of water transport and the use of water for power production, because these needs collided. During the Ming and Qing dynasties, the Director General of Waterways was responsible for management of the Grand Canal and the lower Yellow River basin, and his main task was, and this is significant, to prevent damage to the impressive Grand Canal. This very powerful state agency has been described as an adjunct of the Grain Transport Administration, and the staff there was therefore not interested in building canals for transport of other goods from other places, or in building dams and ponds for mechanical water power production, also because such projects would have made management of the Grand Canal even more complex and difficult.

In England, management traditions and ideas about how water should be used were much more pluralistic and neither the state nor the financial elite had strong vested interests in a particular type of water management of the past. There were of course conflicts between different water usages also in England, but since irrigation and irrigation canals were of marginal importance, the conflicts arose more within the modernising sectors of the economy – between those who wanted to develop the waterways for transport and those who wanted to develop them for power production. Since the waterscape was so diverse and the rivers had been of marginal economic and political importance for other sectors of society, water entrepreneurs could emerge and operate quite unhindered by state intervention and, in fact, usually supported by the state.

The issue of cultural constructions of water was also important and requires more research to be understood. The transformation of the rivers into a form of natural capital and a source of profit for individual entrepreneurs, as happened in England, did not occur in India, where the cultural construction of water or water resources was not concerned with how to put them to work in the construction of a new social and economic order. On the contrary, water had been recognised as a primordial spiritual symbol since Vedic times (Baartmans 1990), and the seven major rivers, especially the Ganges, were pivots of regional sacred geography. Thus, all over the country temples, holy tanks and important places of ritual were to be found on the banks of rivers, a cultural practice often merged with economic activities that most likely influenced the early drive for water engineering works for purely economic purposes.

The climatic history of these countries in the eighteenth century also needs to be studied in much greater detail. For example, during the very same decades in which England prospered, terrible floods and droughts devastated parts of both India and China.[18] These rapid and partly uncontrollable changes in the behaviour of their main rivers obviously contributed to dynastic decline, but nothing has yet been published on how this affected the economic development of these during these decades. Water systems also had a fundamental impact on health, living standards, population growth and thus labour productivity and labour costs. In India, for example, malaria was rampant, and it has been called 'Killer no 1' in the sub-continent. Nobody has so far presented a systematic comparative study of the consequences of waterborne diseases on productivity levels in the eighteenth century.

With its focus on the period from the 1760s to the 1820s, this chapter has argued for the fruitfulness of a broader research project on the Industrial Revolution that would locate it within a deeper history of technological change, waterscapes and water control.[19] There is general agreement that the mixture of large rivers, fertile soil created by the same rivers and irrigation was a precondition for the first riverine irrigation civilisations to develop in Asia and the Middle East about 5,000 years ago. The erratic character of the rainfall and the intense seasonality and silt loads of the rivers were turned into advantages by human intervention. Large and prosperous agricultural civilisations were established with the aid of artificial irrigation. From the Indus to the Yellow Rivers, great civilisations rose from the life-giving waters of these turbid, violent river systems and, especially in China, there emerged a strong water management tradition, geared to control and defence. This chapter has suggested that the very conditions that had given Asian irrigation economies their comparative advantage for millennia had become a disadvantage by the eighteenth century. The system that was conducive to agriculture made it extremely difficult to establish modern factories based on water power along British lines, and stifled the establishment of an efficient system of all-year waterways for transport.

Many factors must be considered in explanations of why parts of north-western Europe were at a disadvantage compared to Asian and Middle Eastern irrigation civilisations during the long period when agriculture dominated. But these different development trajectories cannot be grasped without bringing into the analytical picture the fact that rain-fed agriculture was far from being as productive as cropping benefiting from artificial irrigation and fertile silt-laden soils. The long, snowy winters additionally reduced the cropping period and the possibility of multiple harvests, making overall yield much less in north-westen Europe than on the banks of the Nile, the Indus, the Yellow River or the Euphrates. The Eurasian raincoast states did also develop a water management bureaucracy, but it was much smaller and less powerful than in the river basins further east and in the eastern Eurasion monsoon states. Moreover, here the habits of thought or the ideas about water were not geared to exploiting rivers for canalisation and agriculture, but had for centuries been concerned with how to handle conflicts between millers or transporters, or between the mill and the lock.

Then, in the later eighteenth century, during the first phase of the Industrial Revolution, those hydrological conditions, which had worked against radical productivity gains in the earlier phases of the agricultural age, gave this part of Europe a decisive advantage over Eurasia. The waterscape of much of England, and to a lesser extent of parts of Western Europe, was relatively easy to exploit, control and develop into transport routes and sources of power, thanks to its abundance of medium-sized, perennial rivers and brooks, which could be exploited all year round. The rapid economic development of the latter part of the eighteenth century had been in the making for generations, as thousands of millers, engineers and boatmen had experimented with and improved technologies and machinery capable of exploiting countless small, silt-free, benign brooks and streams. Even if the Chinese and the Ottomans had had the same entrepreneurial ideas and political culture as some of the British modernisers, they could not have pushed their country through an industrial revolution at that time because they did not have either the transport routes or the power sources. This chapter has shown that in order to give a more complete and thorough understanding of why Britain and the West could lead the world into one of its most fundamental transformation processes one also has to make variations in water-society relations a central part of the whole story, and that it is not sufficient to focus only on European exploitation and unequal trade relationships or on institutional traditions and cultural and scientific ideas. By including both social and natural factors in the narrative, this non-reductionist analysis of the Industrial Revolution helps to explain the gradual and regional character of its development and the emergence of the 'Great Divergence', and it does so by distinguishing between necessary and sufficient causes and between causes and prerequisites.

3

RIVERS AND EMPIRE

This chapter presents a comprehensive alternative to the interpretations of the partition of Africa and the Nile basin that have dominated historical literature for a century. Mainstream diplomatic history has regarded European rivalry as the main reason for Partition, concluding that Britain ended up as an imperial power in the Nile basin because London feared French and other European actors' activity there. The British established their Nile empire more or less by default; they did not plan for it but were forced to move up the Nile because of the fear created by France's activity in the region. Many books have also been written arguing that the British march up the Nile was primarily religiously motivated; it should be interpreted by and large as London's revenge for the brutal murder of the Christian martyr Charles Gordon in Khartoum in 1885 by Islamic fanatics. The Sudan campaign is from this perspective seen simply as a forerunner to current conflicts between Islamism and the West. It has also been used as a case proving theories about the inherent imperial drive of modern capitalism. Finally, a more abstract model explains European colonialism in general as a hunt for African riches or as a form of capital export in disguise.

This source-based narrative will argue that all these explanations miss out the most crucial factor in British strategy in north-east Africa: they overlook the fundamental importance of the way the River Nile runs in the African landscape and how geographical and hydrological factors and specific water-society systems influenced and framed British strategies and actions. The combination of the particular confluences between river and society and the hydraulic design for the Nile implemented by the British imperialists must be understood. Appreciating the importance of the above factors does not imply paying less attention to traditional diplomatic correspondence or to the role of individual missionaries or imperial entrepreneurs, but it begs for analyses that are more inclusive, more empirical and more interested in the 'places' and hydro-geopolitical contexts in which policies are implemented and in how the actors interpreted these places.

WATER IMPERIALISM ON THE NILE

In the voluminous literature on the partition of Africa and the role of the Nile quest in this context, European rivalry has by and large been interpreted as a necessary and sufficient cause of British expansionism in the Nile basin.[1] The most influential theory has been that suggested by Robinson and Gallagher in their famous and very influential study of Victorian imperialism. The occupation of Egypt began a domino process which led Britain into Uganda, Kenya and ultimately to the conquest of the Sudan. Britain was primarily concerned to prevent other European powers, particularly Germany and France, from muscling into London's spheres of interest. As a result of this fear or concern, the argument goes, during the last 10 years of the nineteenth century, Britain occupied or annexed the Sudan and Uganda. The prime object was defensive: the prevention of serious inroads being made on British power. Britain was not an instigator of the scramble for Africa, but responded to the actions of other forces. The British were not really interested in the areas where they took control; they became an imperial power there without really wanting it or without having a plan for what to do with their new territories.

In reality, however, and this chapter will show that this was the case, British well-considered Nile imperialism shows that the British advances into tropical Africa from Egypt were informed by a kind of expansionism that went far beyond fear of European rivals or conventional commercial expansion. Its aims were long-term and strategic, indeed, and were essentially related to taking control of the Nile as the region's defining water resource to control it for political means and to further British strategic and economic interests in Egypt, the country harbouring the Suez Canal. One might argue that what are here called London's 'hydraulic calculations' made a British Nile empire a rational strategy. The partition therefore did not accompany, but rather preceded, the invasion of tropical Africa by the trader and the official. It was not regional commercial expansion that required the extension of territorial claims, as one school of interpretation of colonialism has argued is typical for imperial policies, but the extension of territorial claims which in due course required commercial expansion. African resistance and collaboration did not to any important extent influence the direction of the partition of East Africa and the Nile basin. Rather, it was the nature of the Nile itself that conditioned the direction British imperial expansion took and it was its hydrology that particularly influenced the build-up of imperial institutions and imperial organisation.

This chapter argues that although the European 'fear-factor' was an element in British diplomacy in the region, it cannot explain the continuities and shifts in British Sudan policy in the 1880s and 1890s, nor make intelligible the historical documents that clearly demonstrate the existence of persistent plans for British hydro-imperialism in the Nile valley, regardless of what other European states were up to. It was primarily the combined impact of the importance and potential of Egypt's irrigation economy under British leadership after 1882, the repercussions of a growing water crisis in Egypt in the late 1880s and 1890s, and the structuring character of the regional water system that shaped the destiny of the Sudan and the rest of the Upper Nile in the late 1890s.[2]

While it has usually been claimed that the *'frontiers of fear'* on the move motivated the British march upstream,[3] this chapter argues that the *limited irrigation water* available in Egypt on the one hand and the abundance of Nile waters upstream of Aswan waiting to be controlled for the benefit of Egypt and cotton cultivation on the other, made expansion up the Nile a very rational and, as will be documented, a well-planned imperial strategy. This chapter will argue that the Southern Sudan was regarded by the British strategists as a *barrel filled with water* and not as the 'bottom of the barrel', as this region has frequently been described in the historical literature on the occupation.[4] This region possessed huge amounts of water that could very profitably be used to enlarge cotton farms in Egypt, and the waters of the White Nile running through one of the biggest wetlands in the world were described by leading water planners as being as valuable as gold.[5] The Sudan was the very key to the planned development of Egypt and its cotton industry, due to its geographical location in the Nile basin – this was the country where the two Niles meet. To conceive and portray the Sudan as a 'buffer state' between European rivals is therefore for this reason misleading (Robinson and Gallagher 1981: 475). The British had grandiose schemes for controlling the waters of the longest and most famous river in the world with the most modern technology available at the time. These plans were not grounded in a 'defensive psychology' but in feelings of imperial strength and modernising confidence. Instead of a theory that 'suggests the kind of defensive imperialism that extends beyond the areas of expanding economy but acts for their strategic protection' (Robinson and Gallagher, 1981: 474–5), this analysis suggests that the British Nile policy was a kind of promethean hydro-political river imperialism, and an imperialism that extended beyond the area of an expanding economy and that acted for Egypt's continued agricultural and economic development by exploiting the region's most important resource and geopolitical factor.

In order to understand why the British were so obsessed with control of the Nile, why they were more interested in the modest White Nile than in the mighty Blue Nile, or why they thought it was just a question of time before they had to occupy the Sudan, we need to understand the character of the regional water situation. Egypt, lying well downstream on a river running through three climatic zones and what are now 10 countries before it reaches its borders, was basically a desert. About 97 per cent of the people lived along the banks of the river, and the economy was totally dependent on river water and river control in some form or another. Egypt had been the granary of the Roman Empire and was now the enormous cotton farm of the British Empire. The need to even out the annual and especially the seasonal variations of river discharges was obvious, since about two-thirds of the entire annual flow flushed down during some three months in autumn while the last third reached the dry lands during the remaining nine months. The profitable cotton season was in summer, when the natural river flow was at its lowest. Almost all the water, about 80 per cent, during this *sefi* season came from the White Nile, mostly thanks to the natural storage provided by the *Sudd*, the immense swamps of the Southern Sudan and the relatively regular outflow of the Central African Lakes. The British were aware of these factors due to the findings of previous hydrological studies, and thus understood that the White Nile, or the *Bahr al-Jabal* and *Bahr al-Zaraf*, as the tributaries that formed the White Nile south of Kosti were called in Arabic, was the most important tributary. The Blue Nile and the other Ethiopian tributaries contributed more than 80 per cent of the average annual flow at Aswan, but at that time it was not technologically possible to tame them, mostly due to the huge amount of silt in the floodwater which would threaten to destroy the reservoir and sluice gates. Since the watering of the fields in Egypt and the flood security and economic stability of the country depended on control of the Nile, few questions were of greater importance to the British administration than knowledge of how much water would flow into Egypt at any time of the year before it reached the border and the plots of the *fellahin*.

The water-system approach's emphasis on human modification of waterscapes and thus on water management on the Nile focuses on a whole new class of historical sources and documents. Archives in Khartoum, Durham, Cairo and London hold a vast number of records that deal with the Nile and Nile control. The annual reports written by Lord Cromer, Her Majesty's Agent in Egypt between 1883 and 1907 and the 'puppet-master' of Egyptian politics during a couple of decades, letters and minutes of discussions between Cromer and the Foreign Office and ministers in London, the private papers of the leading British actors in the Nile valley living along the river from Alexandra to Entebbe and the extensive Nile discourse and number of Nile plans and Nile actions all clearly show that the Nile campaign of the 1890s was no step in the dark. The British strategists were knowledgeable about the hydrology and geology of the Nile valley, and their expansionist policies were driven by a complex mixture of economic and political considerations, basically influenced by the structuring capabilities of the Nile's geographical, physical and hydrological characteristics which they understood.

'THE EGYPTIAN QUESTION IS THE IRRIGATION QUESTION'

By the early 1890s, the Government in London and Lord Cromer and his administration in Egypt had for some years already understood the consequences of governing a hydraulic society whose development was totally dependent on the waters of the Nile. Every ruler of Egypt had found that the provision of sufficient water for irrigation had been fundamental in achieving political stability and economic prosperity. The British realised that security and stability hinged on their ability to develop the Nile. Egypt had also become more and more important as a source of cotton for the still crucial Lancashire textile industry, partly because of the repercussions of the American Civil War, and partly because of the high quality of the cotton produced in the 'land of the Nile'. The then Egyptian prime minister Nubar Pasha (1884–8 and 1894–5) summarised the situation in a famous one-liner: 'The Egyptian question is the irrigation question' (quoted in Willcocks 1936: 67). Words and deeds show that the British concurred and understood the implications of this Nile dependency; they knew that a downstream hydraulic state on one of the longest rivers in the world simply had to develop its life artery, both in Egypt itself and upstream of its borders in order to survive and thrive.

The British administration under Lord Cromer understood the importance of the Nile from the very beginning and soon they also acquired quite good knowledge about the workings of its hydrological system. One of the first actions Lord Cromer took was to bring experienced water engineers from India, where river control works had long been a priority of the British administration. The demand for more summer water in the 1880s and early 1890s was heard from all levels of Egyptian society, as well as from influential pressure groups, such as the cotton lobby, in Britain. In Egypt, the most powerful foreign trade agencies dealt in cotton (Tignor 1966: 234). The big landowners owned about two-thirds of the cotton harvest. The population doubled during a few decades and reached almost 10 million in 1897, and the growing number of poor peasants put pressure on the Government to provide more reliable water supplies. In England, the Lancashire cotton industry was searching for ways to reduce their dependence on American cotton, and imports of cheaper but very good-quality cotton from Egypt became more and more important. Furthermore, British banks had a strong and growing interest in a thriving Egyptian economy, mainly because in 1882 Egypt's foreign debt had risen to £100 million, and annual debt servicing came to £5 million (Crouchley 1938: 145), much of which went to Britain. Egypt's ability to repay its debts to British banks depended to a large extent on cotton exports and the value of agricultural land. A telling contemporary reflection of this 'Nile water awareness' in London was the fact that *The Times* reported regularly on the water discharges of the Nile! Thus the general political and economic development and the changes in the world trade patterns of cotton led to mounting pressure on the British rulers in Cairo to provide more water to the fertile but dry lands along the banks of the Nile.

The British had barely planted their flag on the banks of the Nile before they were met by vigorous demands for large hydraulic enterprises (Scott-Moncrieff 1895: 414–15). With a growing demand for water on the one hand, and a river far from being harnessed on the other, any administration in Egypt in the late nineteenth century would have been obliged to make increased water control a top priority. The overarching questions became: How to increase the Nile yield in the 'timely season', that is, during the summer season, when cotton was grown and the natural discharge was at its lowest? How to protect agricultural lands against devastating floods? How to dam the excess water in September, October and November for utilisation in the season of water scarcity? How to construct dams which could reduce annual discharge fluctuations? How to narrow the gap between the availability and demand for water was a permanent source of worry to the British. The complexities of this task increased as perennial irrigation spread and demonstrated its economic potential. As the British faced rising expectations, their legitimacy as rulers of an irrigation society required that they should succeed in narrowing the growing gap between supply and demand. At first, the hydraulic engineers concentrated on what could be done by improving existing irrigation facilities and building some new water-control structures within the borders of Egypt.

BRITAIN AND THE NILE IN EGYPT

Thanks to a revolution in irrigation methods, Egyptian agriculture had undergone important transformations in the decades prior to the British invasion in 1882. The old system of flood-irrigation had been replaced by all-year irrigation. Perennial irrigation on a larger scale had started under Mohammed Ali, who developed an agricultural strategy based on an assessment of Egypt as having the perfect climate, fertile soil and an abundance of people; the problem being the lack of water. In 1820, cotton production and exports were negligible, whereas after the delta barrages had been built and new canals dug in the middle of the nineteenth century, cotton made up about 80 per cent of Egypt's total exports from the 1860s onwards. For a number of reasons, these waterworks fell into disrepair in the following decades, and they were further damaged during the failed nationalist rebellion against the British. Not until after the occupation, and under Cromer's watchful eye, did the priority become repair and improvement of the existing irrigation system (see, for example, Scott-Moncrieff 1895; Willcocks 1889 and Willcocks and Craig 1913).

A series of important though smaller projects included the remodelling of the Upper Egypt basin, clearing silt deposits from the canals and starting operations at the Mex Pumping Station in Lower Egypt. Altogether, these works, more efficient management of the irrigation sector and an improved system of drainage and crop rotation contributed to the doubling of cotton production from 1888 to 1892 (Crouchley 1938: 148). In 1891, the British repaired and made operable the delta-barrage system just north of Cairo. It extended the area on which cotton could be grown and it reduced the amount of labour required to bring water to the field. Perennial irrigation was now possible over the entire cultivated area of the Delta. This brought great material advantage to Egypt, and it also led to the abolition of the *corvée*.[6] As long as this work within the borders of Egypt was a priority of the Ministry of Public Works, and the Government was in grave financial difficulties, there was neither capacity nor any need to look upstream of Egypt for a more efficient way of using the Nile waters.

By the early 1890s, however, the upper limit for expansion within the existing Nile control system had been reached. The natural character of the annual and seasonal discharge fluctuations and the human efforts to modify and control the way the Nile waters ran in the fields after the floods had created an irrigation system that was clearly insufficient for the current and growing needs of water. In spite of British efforts to improve it, it did not even always satisfy actual demand, with grave economic consequences for the cotton industry. In 1888, for instance, about 250,000 acres in Upper Egypt received no irrigation water whatsoever.[7] The irrigation officers reported to Cromer in that year that the spirit of resistance to the British presence was 'stronger now than ever'.[8] In other years the seasonal autumn flood was bigger and more violent than usual, and caused great damage to the harvest and the economy in general: the British flood control system was in the 1890s not greatly different from what it had been for centuries.

The combination of the great potential of the irrigation economy and ambitious plans for more profitable cotton farms on the one hand

and the combined danger of devastating floods and the actual growing water gap during the summer season on the other hand, asked for more revolutionary initiatives and developments in water control. In the early 1890s, a modernising Egyptian Nile discourse developed; speeches were made and plans put forward and debated, reflecting the growing ambitions and a feeling of a deepening water crisis in Egypt. J. C. P. Ross, former Inspector-General of the Egyptian Irrigation Service, summarised this attitude, when he wrote in 1893: 'We have now arrived at a stage in the summer irrigation of Egypt where the available natural supply has been completely exhausted, and there still remains more land to grow cotton' (Ross 1893: 188). Both 1889 and 1890 had experienced an exceptionally bad summer supply due to low natural river discharges, immediately causing great falls in profits and increasing the danger of political unrest. Waterworks of an altogether new type and technology were required, and were considered. It became increasingly evident that the age-old system of flood irrigation, or basin irrigation, and the primitive system of summer irrigation, basically being a clever adaptation to the seasonal fluctuations of the Nile, had become inadequate, and that these hydrological fluctuations would have to be controlled and evened out. Scott-Moncrieff, the Under-secretary of the Egyptian Ministry of Public Works, decided that a detailed study of reservoir sites should be a top priority. In 1894, the *Report on Perennial Irrigation and Flood Protection of Egypt* was published by the Government, after having been confidentially circulated in 1893.[9] It estimated the future annual need for summer water at 3,610,000 cubic metres (Willcocks 1893: 9). It asserted that if irrigation were introduced in Upper Egypt, where agriculture still depended on the basin system, and improved in Lower Egypt, the annual income would rise from £E(Egyptian pounds)32,315,000 to £E38,540,000.[10] This report, that was produced the year before the occupation of Uganda in 1894, posed the overarching political and administrative question: How to secure over 3.5 billion cubic metres of irrigation water in the summer season, creating an estimated annual net gain of £6,225,000 for the country? And, at the same time, how could the country be defended against devastating floods?

The most concrete suggestion contained in the 1894 report was to build the reservoir at Aswan in Upper Egypt that had already been discussed by the Government. However, this reservoir, in spite of the fact that it would have been by far the biggest in the world at the time, was seen as a temporary solution only, because the planned capacity satisfied only half of Egypt's estimated needs. At a meeting on 3 June 1894, for example, the Egyptian Council of Minsters discussed other possible dam sites, but this time in the Sudan as if building dams on the river in another country was no noticeable obstacle.[11] In line with this, in the same year Cromer wrote that the Aswan Dam within Egypt's borders may 'at some future time, [...]

perhaps be supplemented by another dam south of Wady Halfa',[12] that is, in the Sudan. And William Garstin, Cromer's right-hand man, underlined in his annual report (1894) that the 'construction of a second [...] [dam] [...] to the south will be merely a question of time'.[13] He further wrote that 'we may confidently predict' that the Egyptian dam will be 'only one of a chain which will eventually extend from the First Cataract to the junction of the White and Blue Niles'.[14] William Willcocks, the main architect behind the dam, stated what for the water planners would have been obvious: that the 'infinitely better and more reliable' flood protection for Egypt was to 'control the Nile before it enters Egypt' (Willcocks 1894: 45). In the early 1890s, the British administration in Egypt thus discussed again and again the need for Egypt to modify and take control of the Nile upstream of the old borders of the country.

This 'chain' of water works upstream became a much more pressing issue when the administration realised that the planned storage capacity of the Aswan Dam, 2,550,000,000 cubic metres of water, would have to be drastically reduced due to technical difficulties in damming the silt-laden floodwaters. After more careful studies of the hydrology of the river and the deposits it carried from the mountains of Ethiopia, London realised that they could just store what they called the 'tail-end' of the flood. Additionally, unexpected political problems arose. In autumn 1894, just after publication of the new report,[15] archaeological groups in France and Great Britain united in demanding a lower than planned water level in the dam, in order to save the ancient temple at Philae close to Aswan from inundation (Scott-Moncrieff 1895: 417). The conflict of interest between those wanting to save the old temple erected on the banks of the river, where it flows into Egypt from the Nubian desert, for the heroes of the ancient Nile cult, the fertility gods Isis and Osiris, and those who wanted to subjugate the Nile for the needs of economic growth collided. The opposition was so strong that it forced the Government in Cairo to yield and to amend its 1894 plan. According to Garstin, the capacity was therefore reduced by more than 50 per cent, to 1,065,000,000 cubic metres.[16] The reservoir could therefore meet only 25 per cent of Egypt's future needs.[17]

According to Garstin, the reduction meant that 2.610 billion cubic metres would have to be found elsewhere.[18] Of course, for geographical and hydrological reasons, this 'elsewhere' could not be along the Nile in Egypt, first and foremost because of the silt which the Blue Nile and Atbara carried from the Ethiopian highlands. The sedimentation problem also excluded 'any hope of constructing solid dams of the ordinary type in the valley of the Nile downstream of the Atbara junction' (Willcocks 1894: 12). The Nile could thus, it was thought at the time, only be profitably dammed and controlled upstream. The reduction in the size and storage capacity of the Aswan Dam made the question of upstream expansion and water control there a much more pressing issue. Now military and political expansion upstream had clearly become a strategic issue of central importance to London.

BRITAIN ON THE UPPER NILE

From the very beginning, the water planners knew that an Aswan Dam within the borders of Egypt could not be operated rationally without better and more accurate knowledge of the Nile, upstream and in the Sudan. Precise hydrological information about the fluctuations in the flow of the tributaries before the main Nile reached the reservoir was essential. In 1894, Willcocks showed that the time the waters took between Khartoum and Aswan was only '10 days in flood and between Aswan and Cairo only five days'. Obviously, proper management of the reservoir and its gates – especially since it only needed to store the tail-end of the floodwaters – therefore required a number of gauging-stations along the Nile and its tributaries in the Sudan, as well as the re-establishment of a working Nilometre in Khartoum at the junction of the Blue and White Niles. As early as 1882, before the era of reservoirs, Major Mason Bey had shown the necessity for establishing more Nilometres on both the main Nile and its tributaries in the Sudan for planning purposes in Egypt (Mason Bey 1881). In May 1893, the Societé Khédival de Géographie discussed in detail the water discharge information gathered from the gauging-stations in Sudan, established on the order of Ismail, from the time when, as they expressed it, 'the Sudan was not closed' (Ventre Bey 1883). Until 1885, Egypt had daily received information by telegraph from the Nilometre at Khartoum,[19] and in 1875 a station was erected close to the village of Dakla in order to measure the Atbara (Chélu 1891: 35). The 'fall of Gordon' in 1885 was dramatic and caught the attention of the day (and of historians later on), but the loss of the Nilometre at Khartoum represented a more direct threat to Egypt, because it jeopardised the optimal management of the irrigation system.[20] However, what the water planners in Cairo regarded as a great loss as early as 1885 had far greater consequences in the mid-1890s as a result of the growing water gap, the vulnerability of the new crop-rotation system and the more exact hydrological information required for the planned big reservoirs. Willcocks wrote in 1893: 'As Egypt possesses no barometric, thermometric, or rain gauge stations in the valley of the Nile, we are always ignorant of the coming flood' (Willcocks 1893: 17).

At that time, the British hydrologists and engineers lacked in-depth knowledge of the Nile's upper reaches. Ross wrote that 'unfortunately the Dervishes prevent any scientific examination' of the Nile upstream

(Ross 1893: 191). Scott Moncrieff complained, while lecturing in London in 1895, that he, like his audience, had to go to 'the works of Speke, Baker, Stanley and our other great explorers' for information regarding anything higher upstream than Philae, and said that 'if a foreigner were to lecture to his countrymen about the river Thames, and were to begin by informing them that he had never been above Greenwich, he might be looked upon as an imposter' (Scott-Moncrieff 1895: 405). William Garstin described these years, as far as hydrological studies were concerned, as if a 'thick veil had settled down on the Upper Nile' (Garstin 1909: 135).[21]

Putting the Nile and the need to control it in the centre of the picture opens up a whole new class of historical sources dealing with Nile planning and Nile management, and it becomes evident that several years before the Sudan Campaign started, like Scott-Moncrieff, Ross, Willcocks and Garstin, Cromer's water engineers in charge of the Ministry of Public Works were discussing the necessity of controlling the Nile upstream of Egypt. A central vision in the government report of 1894 was that the hydrological features of the Nile and the future increase in summer water demand would require the regulation of the Nile south of Egypt, even as far as Lake Albert and Lake Victoria. Willcocks wrote that what 'the Italian Lakes are to the plains of Lombardy, Lake Albert is to the land of Egypt' (Willcocks 1894, Appendix III: 11). By damming the lake(s), 'a constant and plentiful supply of water to the Nile valley during the summer months' could be ensured (Willcocks ibid., Appendix III: 11). 'There alone,' he wrote, 'we deal with quantities of water which approach' the demand (Willcocks ibid., Appendix III: 10). The previous year Ross had speculated along similar lines. He envisaged that raising the water level of Lake Victoria by only 1 metre would bring a flow that would be '30 times more than wanted' (Ross 1893: 189). In 1894, London took military control over the African lakes that several decades ago had been 'discovered' by Speke, Burton and Baker and named by them Lake Victoria and Lake Albert (Tvedt 2004a: 19–51). No administration in Cairo would ever consider regulating Lake Victoria, which is roughly the size of Scotland, without controlling the shores of the lake, and without improving the White Nile's water transport capacity in Southern Sudan, due to the river's natural water losses there. Garstin and Willcocks knew that *sadd* was blocking the river,[22] and that the White Nile lost huge amounts of water on its way through the swamps in Southern Sudan.[23] They knew very well that it would be impossible to improve the knowledge of the Nile unless the river was cleared of *sadd*.

In the Southern Sudan the British were therefore not scraping the 'bottom of the barrel' (Robinson and Gallagher 1953: 15). This part of the Nile basin was, on the contrary, filled with extremely valuable summer water. Because of its waterscape and hydrological characteristics, this area

55

needed to be controlled, according to Cromer and his water experts, both politically and water management-wise, from London and Cairo. London was not preparing for war for 'the mastery of these "deserts"', as this area was totally misconceived by the most influential historians on the whole issue of the Partition of Africa (Robinson and Gallagher 1981: 372), but because this area held (and still holds, by the way) the future of Egypt in its hands, or more precisely, in its swamps. The occupation of the Southern Sudan was therefore not 'an imperialism without impetus' (Robinson and Gallagher 1981: 25), fighting for an 'illusion' or an 'empty barrel' but an imperialism with a very, very strong impetus and rationality. Planning for the optimal usage of the Nile waters inspired thoughts about the Nile as a single river basin that should be under one political planning authority, and this plan for conquering an unruly waterscape was a motivation of its own. The sheer magnitude of the task made the water planners compare themselves with the already famous British names in Nile history. The discovery of the sources of the Nile had brought fame to their countrymen Speke, Grant and Baker. Now Garstin, Scott-Moncrieff and Willcocks could follow in their footsteps, they could even 'take the river in hand'.[24] In 1894, Willcocks directly described their plans for the Nile as a worthy follow-up to these British discoveries. Garstin later wrote that if they succeeded in taming the Nile such an accomplishment could be compared with the building of the pyramids (Garstin 1904: 166). In 1895, Scott-Moncrieff summed up the 'Nile vision' of the water planners when he said: 'Is it not evident, then, that the Nile from the Victoria Nyanza to the Mediterranean should be under one rule?' (Scott-Moncrieff 1895: 418)

BRITAIN AS THE LORD OF THE NILE

In his *Modern Egypt*, Cromer wrote that a central motive for the occupation of the Sudan had been 'the effective control of the waters of the Nile from the Equatorial Lakes to the sea' (Cromer 1908, II: 110). Full of confidence in his imperial Nile strategy, he wrote: 'When, eventually, the waters of the Nile, from the Lakes to the sea, are brought fully under control, it will be possible to boast that Man, in this case the Englishman, has turned the gifts of Nature to the best possible advantage.' (Cromer 1908, II: 461)

The first decades of British rule on the Nile were termed by later irrigation advisers the 'Cromer-Garstin regime',[25] a regime under which the most powerful politician and the most powerful water planner developed a consistent and overall strategy and a plan for Britain as a River Nile empire. Garstin's department was given an exceptional degree of autonomy and was deliberately shielded from intervention by other European interests in Cairo, and was staffed with a number of British experts.[26] Cromer later wrote that these expenses 'contributed probably more than any one cause to the comparative prosperity of Egypt' (Cromer 1908, II: 464), and ensured no less than 'the solvency of the Egyptian Treasury' (Cromer 1908, II: 464). According to Cromer, irrigation works were not only a permanent priority, but also a policy which continuously proved its success.[27] From 1890 onwards, every Annual Report to the Government in London enclosed a separate Memorandum on irrigation activities. Everybody seemed to agree: 'The best thing the Financial Ministry can do is to place as much money as it can afford at their disposal [British water planners, my comment], confident that whatever is thus spent will bring in a splendid return.' (Milner 1892: 310)

Summing up British rule from 1882 to 1907, Cromer put hydraulic engineers on an equal footing with the army for internal political reasons; they created the situation that made Egypt and Suez safe for the British. While the soldiers held the Egyptians down by force, the water planners won their minds, or as his financial adviser put it in 1892: the British engineers secured the support of Egyptian public opinion (Milner 1892: 310). They 'justified Western methods to Eastern minds,' Cromer wrote (Cromer 1908, II: 465). Or as he had already put it in 1886: 'the good results of European administration can readily be brought home to the natives' (quoted in Zetland 1932: 171). Two years later, he wrote that British success in Egypt depended on development of the irrigation structure and increased access to summer water.

It was not only the men in charge of Nile development who in the early 1890s were discussing control works on the river in the Sudan. In 1891, Cromer wrote a long letter to Prime Minister Salisbury in London on the reservoir question. He said that all competent authorities agreed that something had to be done, but not on what had to be done. He discussed different options; the reservoir might be constructed 'either at Wadi Halfa, or at Kalabalah, or at Assuan, or at Silsileh, or a reservoir might be made in the Wady Raian'.[28] He said that the subject was one of 'utmost importance', because 'the prosperity of Egypt depends wholly on the Nile'.[29] In November 1891, Cromer again told Salisbury about the importance of the water storage question in Egyptian public opinion.[30] In 1893, he sent a telegram to Lord Rosebery, foreign secretary in the Gladstone Government, supporting a circular which had been addressed to the Powers by the Government of His Majesty the Khedive, requesting that the economies 'effected by the conversion of the Debt should be applied to the constructing of reservoirs in Upper Egypt'.[31] He supported the 1894 report and not only actively backed the plan for the Aswan Dam but was very active in securing money and political backing for its construction.

As long as it was not clear whether the British were going to stay in Egypt, and as long as Egypt had enough water for their summer cultivation, and had no money to finance either reservoirs or wars, Cromer and the London Government rejected more adventurous proposals to march southwards. Cromer informed London that he disagreed strongly with those who in the 1880s wanted to occupy the Sudan. In 1884, he asked whether the British Government intended to establish a settled form of government at Khartoum or not and he answered himself 'in the negative'.[32] If the aim was to have slavery completely abolished in the Sudan, a small expeditionary force would not be enough, he argued. He ridiculed those in England who publicly argued in favour of such a policy, by stating that then 'you must send an English army to occupy the country',[33] which nobody was prepared to do. As late as 1886, he wrote to London, saying that all the authorities in Cairo except himself were in favour of an advance to Dongola. He was 'opposed to making any advance at all', while the Egyptian authorities, he argued, favoured the idea because they regarded it as 'a first step towards the reconquest of the Soudan'.[34] London agreed with Cromer's reasoning. It seems clear that both the British Government and Cromer were looking for an opportunity which could legitimise the occupation in both Egyptian and British public opinion, and that they objected to imperial adventurism but favoured an occupation that could be sustained.

Just before 1890, there are clear evidences that Cromer changed his rhetoric. Now he wrote about the occupation as being necessary – one day – while still arguing in favour of playing safe and only acting when time was ripe and the moment right. In 1890, the British military discussed the occupation of the Sudan. There was general agreement that Dongola in northern Sudan 'from a purely military point of view, could only be of use to us as a stepping stone, as an advanced base for an advance upon Berber or Khartoum'.[35]

The way Cromer and the British Government connected hydrological research and water planning to the military campaign clearly shows that concerns and long-term strategy were based on a deep understanding of Nile politics. Some months before the British occupied the Nile upstream in 1898 and Cromer sent his most senior water planners in their wake all the way up to Lake Victoria and Lake Tana, he wrote to Prime Minister Salisbury: 'There can be no doubt that the *most crying want of the country* [my italics] at present is an increase in the water supply.'[36] No sooner had the British moved into the Sudan than he sent – in his own view – his most important official in Egypt on an expedition up the Nile. In April 1897, Garstin had submitted his report on the Nile cataracts.[37] In the wake of Kitchener's flotilla, Garstin studied the White Nile in 1899, the White Nile, Bahr al-Jabal, Bahr al-Zaraf and Bahr al-Ghazal in 1901 and again in 1904. In 1903, he was in Uganda, along the Semliki River, at Lake Albert and again at Bahr al-Jabal (Gleichen l905: Vol. 1, 280). When

Garstin in 1899 proposed to remove the *sadd* in the Bahr al-Jabal which blocked the river's flow, he received immediate financial support from Cromer. Cromer's argument was: 'The question of increasing the summer supply of the Nile is, however, of such a vital interest to Egypt, that the present expenditure is fully justified.'[38]

In his introduction to Garstin's 1904 report, Cromer gave priority to the plans on the Upper Nile. Cromer suggested that £E5.5 million should be allocated for the proposed regulation works in the swamps.[39] The projected cost of the recommended investments is most clearly illustrated by comparing it to the total cost of the Sudan campaigns from 1896 to 1898, that is, £E2,345,345 (Peel 1969 [1904]: 263), and compared to the total revenues of the Sudan budget in the years 1899–1903, that is, £E1,132,000.[40] Of course, Cromer had no intention of using this money, a sum which far exceeded any investment the British had previously made in the Nile Valley, to scrape the 'bottom of the barrel'.

In March 1898, Cromer wrote Salisbury a long letter on the question of the occupation of the Sudan, arguing that he had 'always been fully aware of the desirability of bringing the Soudan back to Egypt'. He even drafted, but then deleted, the following sentence in the final letter: 'I have, therefore, always looked forward' to the occupation of the Sudan. What Cromer awaited was for 'essential conditions' to be in place. He wrote: 'The great mistake made by Ismail Pasha was that before he had learnt to administer efficiently the Delta of the Nile, he endeavoured to extend Egyptian territory to the centre of Africa.' His experience should be a 'warning', which had to be told to and taught to the Egyptians, Cromer wrote.[41] His annual reports and his letters to London show that Cromer now thought that the British had learnt how to administer the Delta, that the economy was sound, and that the demand for more summer water was therefore rising. The moment was approaching.

The plans of Cromer and London were not easy to accomplish. For economic and political reasons they wanted Egyptian rather than British troops to do most of the fighting. Their aim was that the Egyptian Treasury and not the British Treasury should pay the cost of the expedition. It required political competence and diplomatic ability to achieve this aim, not least because there already were legal wrangles over the financing of the Dongola expedition.[42] The Egyptian Government, which was under British control, had demanded the withdrawal of £E500,000 from the general reserve fund. The International Public Debt Commission in Cairo had allowed the withdrawal of funds only by a majority vote. But Cromer and London succeeded in the end: the Sudan campaign was paid for by the Egyptian Treasury and mostly fought by Egyptian soldiers.

How could the occupation over the whole of the Sudan be justified to British, Egyptian and European opinion? London had already 'revenged

Gordon' by taking Khartoum. But why move on to the swamps of Southern Sudan as well? London was looking for the right arguments that could get support from Egypt and win over opposition or indifference in Britain. The French threat was a good diplomatic card, because French imperialists were publicly talking about sending troops to the Upper Nile area. London found the scapegoat they needed in Captain Marchand. In July 1898, after the Dongola war was over, Cromer attended a Cabinet meeting in London to discuss Nile valley policy. Salisbury wrote to the Queen about this meeting:

> The other question [of the Cabinet meeting, my comment] was our dealing with the Nile Valley, if, and when, we had taken Khartoum. For this question Lord Cromer attended the Cabinet and gave us the benefits of his views [...] He thought that the Egyptian and British flags should fly side by side: that the gunboats with Gen. Kitchener and a small force should go up the Nile as far as Fashoda (600 miles): and as much farther as was practicable: and that any other flag in that valley should be moved.[43]

Since Britain's position and military advance depended upon Egyptian support, the sudden appearance of a small group of French soldiers at Fashoda created a golden opportunity: the British could portray themselves as a guardian of Egyptian interests *vis-à-vis* French imperialism and French opposition to the re-conquest. When the French flag went down at Fashoda and the miserable 'force' of Marchand was forced to leave the Nile basin, Kitchener therefore cleverly hoisted not only the British but also the Egyptian flag on the shores of the Upper Nile.

FASHODA AND HYDROLOGY

Many influential historical reconstructions of 'the race to Fashoda' have ascribed to Victor Prompt, a Frenchman working in Egypt, a highly important role in the imperial rivalry in the Nile Valley.[44] His speeches in the early 1890s are said to have created a sort of 'nightmare' among the British rulers, and it was his speculations and the support they obtained in France that made it necessary for Britain to move upstream to stop the French plans for the Nile. How this 'innocent man' was made into a 'villain' in the historical reconstructions of the partition of Africa is an example of potential consequences of misunderstanding or negligence of particular water-society systems and hydrology.

Prompt's ascribed role in the literature about British Nile policies and the Nile Quest of the 1890s is relevant in the context of this chapter, because the way his speeches have been interpreted and misunderstood is related to the same literature's lack of attention to or understanding of the regional hydrology of the Nile. The reality is that Prompt never did say what historians later have claimed, but more importantly, his whole thinking about Nile control has been misinterpreted. His contemporary influences on British policies have furthermore been greatly exaggerated; Prompt caused no stir among the British at the time and he probably played an insignificant role, if any, in directing French imperial Nile policy. In the literature he has been described as a hydrologist, as if he was a Nile expert, but he was an engineer who came to Cairo and Egypt in 1889, appointed as *L'Administrateur français des chemins de fer égyptiens*. It is true that he, like many other people at the time, gave several speeches, on the control and exploitation of the Nile.[45] But the interpretations of Prompt's ideas in the literature have been limited to only one of his four speeches: the 'Soudan Nilotique' from January 1893.

In this speech, Prompt's main agenda was to convince his primarily Egyptian audience that *Egypt* should immediately occupy the Sudan. The reason behind his proposal was clear and conventional at the time: by taking control of the Nile south of her borders, Egypt could secure her water supply. He thought that the flow of the Nile was diminishing due to changes in its *natural* water discharge, or perhaps to natural climate change. This natural alteration in her life-giving river system would have dramatic consequences for Egypt's 'whole existence'.[46]

Egypt ought therefore to occupy the whole of the Nile basin. He proposed three reservoirs between Khartoum and Aswan.[47] He also suggested that the Nile should and could be made navigable up to Khartoum.[48] If implemented, he argued, Egypt's military conquest and occupation of the Sudan from the north would be facilitated. It was a very ambitious proposal that if implemented would have changed the water strategic relationship between Egypt and the Upper Nile for ever. Prompt also suggested building a railway 'de Keneh à Koseir'.[49] He concluded that by urgently implementing these projects,[50] Egypt would benefit from an immense extension of her agricultural area, and would be able to abandon old irrigation methods.[51] In an annexe to one of his papers, he discussed future irrigation projects in the eastern Sudan along the Blue Nile, arguing that Egypt could profit from using excess Blue Nile water there without any negative effects on its water supply during the summer season.

His second speech in 1891 dealt only with reservoirs in Upper Egypt. He said the reservoir question was unquestionably the single most important issue. The speech offered strong support for the planned Aswan reservoir. In January 1983, Prompt again advocated Egyptian reoccupation of what he significantly called her 'lost provinces'. He offered a broad description of the whole basin and suggested how it could best be exploited for the benefit of Egypt. The third part of his speech had the subtitle: 'Intérêts agricoles et commerciaux de l'Égypte dans les contrées que forment le bassin du Nil' (Prompt 1893: 95). Prompt discussed the pros and cons of a barrage at the Equatorial Lakes (Prompt 1893: 72). This speech did no more than discuss the reservoir plans that the British published in the following year. Prompt wrote that, if desired, the Nile could be dammed in Uganda, in order to give Egypt important and much-needed water (Prompt 1893: 101), a project which, according to him, could not be opposed on sound grounds. Prompt did not at all suggest that the French should occupy the Upper Nile and build reservoirs there, as later historians have argued. Neither did he support British ambitions there. What he did was to point to the potential threat to Egypt from a *British* presence on the headwaters of the White Nile (Prompt 1893: 109).

Prompt's speeches supported Egyptian expansionism, and warned about British intentions upstream. He did not play the French card or speculate that France could throttle Egypt at Fashoda, nor did he suggest that either Fashoda or the Bahr al-Ghazal was the hydrological key point in the Nile basin. Prompt never, in fact, mentioned a dam at Fashoda. Contemporary sources did not pay attention to his speeches, because the ideas expressed in them were mainstream in Egypt. A. Silva White's book from 1899, dealing with irrigation in Egypt and the importance of basin-wide water development, did not mention Prompt.[52] Nor did Peel's *Binding the Nile* (1969 [1904]). Cocheris (1903) discusses Prompt, but only in passing. There is no evidence in the sources that anybody at that time regarded Fashoda as a hydrological 'key point' in the Nile valley.[53] Samuel Baker, for example, who was more familiar with this area than any other European (he had been Ismail's governor of the region in the 1860s, with his headquarters at Fashoda), had several times described the extreme flatness of the area, mentioning that the country around Fashoda was 'dead flat' (Baker 1867, I: 44). In none of his much earlier bestselling books is Fashoda even hinted at as a hydrological key point. Shcweinfurth's *The Heart of Africa*, another bestseller, described the area in a similar way, while Lombardini's description (1865) conveyed the same story, and was widely consulted by the British water planners. Emin Pasha published his diaries (1879) concerning the 'Strombarren des Bahr el-Gebel', in which he emphasised the flatness of the area (Emin Pasha 1879: 273). Several much-read French authors, from Arnaud to Chélu (1891), showed beyond doubt that if anybody wanted to dam the Nile, he should definitely not attempt to do so at Fashoda. The assumption that has prevailed in the literature that Fashoda was the key or 'the headwaters' of the Nile, or that contemporary British strategists thought that was the case, is simply wrong.[54] A potential French force at Fashoda therefore did not represent a threat to the Nile flow. It did not create fear in London but would cause an outcry in Egypt, since it struck at the very heart or symbol of their lost Nile valley empire, which they were fighting, under the leadership of Abbas II, to get back.

In the early 1890s, the later Lord Lugar and future British ruler of Uganda wrote: 'Egypt is indebted for her summer supply of water to the Victoria Lake, and a dam built across the river at its outlet from the lake would deprive Egypt of this.' (Lugard 1893, II: 584) And further: the 'occupation of so distant a point as Uganda would be a fair and just claim to render valid our influence in the Nile basin and beyond.' ((Lugard 1893, II: 560) Finally, he quoted Lord Rosebery, who had said that Uganda commanded 'probably the key to Africa' (Lugard 1893, II: 584). In 1894, London took formal and direct control over the African lakes and declared a protectorate over Buganda, in line with Lugard's proposals. They thus indicated that part of this great lake was considered hydro-politically important, since that was where they believed the source of the White Nile to be and the place where a dam could be built. In the same year, they established a Nilometre at the outlet of the lake, and Cromer and his water planners and hydro-politicians could continue working on plans for the entire Nile system.

RIVERS AND EMPIRE

This chapter has rejected the dominant explanation that it was fear of the French or of other European powers that primarily motivated British expansion upstream on the Nile and thus led to the Partition of Africa.[55] During the 1890s, the British developed an ambitious strategy and diplomatic and military tactics for the establishment of a British 'river empire' on the Nile. London had two strategic aims in relation to the Nile upstream of Egypt. On the one hand, to develop the Nile and increase the amount of irrigation water in the desert countries in the north of the basin so as to bolster cotton production and cotton exports to Lancashire, and to improve the Egyptian national economy, since achieving the latter would create political stability in the country and thus also at Suez. London knew also, on the other hand, that control of the Nile upstream would give Britain political leverage against Egyptian nationalists, if need were to arise in the future. The Nile could be used both as a carrot and a stick in order to maintain British control over Suez. For diplomatic reasons the occupation of the Sudan was 'sold' as an Anglo-Egyptian occupation, which also secured the support of the Egyptian elite and funds from the Egyptian Treasury and a supply of soldiers from the Egyptian army. European rivalry on the Nile in the 1890s impacted on British imperial tactics, but this rivalry theory cannot explain British strategy and imperial policies in the Nile basin.

London's and Cromer's grasp of the importance of the Nile and the irrigation question made them fully aware of the fact that if they set foot upstream they would also be able to control a hydraulic state like Egypt politically, and that improved Nile control upstream was necessary in order to give Egypt the summer water that the cotton economy and political stability depended on. The empirical and theoretical argument was that the role of the *sadd* in the Southern Sudan in decreasing the Nile flow during the summer season, the very high sediment loads of the Ethiopian Nile tributaries, and the relative importance of the Blue and the White Nile tributaries and their seasonal fluctuations were all issues of great practical and political concern to the imperial strategists, and such geographical characteristics framed the way in which the imperial strategy was formulated and implemented.

This empirical study of hydrology and empire and imperial hydraulic designs in the context of the Nile and the partition of Africa can stimulate similar research on how other water-society systems impacted on the way European expansion happened in different parts of the world. In the modern period, and given the technological superiority of the European countries at the time when it came to utilising rivers for transport for both military and economic means and for irrigation, river systems affected form, content and motivation for expansion. The role of the Ganges and its tributaries, both in relation to warfare, transport and economic politicies, can hardly be overestimated as factors in British India policy. The same holds for the Zambezi in Southern Africa whose early history is as a highway for the spread of trade and Christianity, symbolised by the missionary of all missionaries, David Livingstone. The Yangtze was at the heart of the relations between Britain and the weakened Qing Dynasty in China, symbolised by the humiliating Treaty of Nanjing in 1842. Any story about French colonialism in Southeast Asia must put the Mekong in the centre of the picture, just as the great rivers of North America helped to structure the colonialisation of that continent. The proposition is that comparative studies of the geographical and hydrological character of these river systems and of how the imperial actors conceived of them and their military, economic and geopolitical importance will bring forth new insights into the role of European imperialism in general. The methodological and analytical strength of the water-system approach is that it enables the location of intentions and acts of historical subjects within specific, modifiable geographical contexts. Empirically, it opens up a whole new set of source material, embedding the reconstruction of different discourses in diversifying water bodies and studying how actors have been influenced by these life-important resources' economic, political and hydrological value.

4

RELIGION AND THE ENIGMA OF WATER

All over the world people have at all times attached a wide variety of religious meanings to water and the permanent uncertainties and flux of the hydrological cycle. Systematic comparisons of the role of water in different religions has therefore a great untapped potential: (a) water is an absolutely essential resource in all societies, (b) most religions give water a central place in texts and rituals, (c) the paradoxical natures of water – it is a life-giver and life-taker, alluring and fearsome, creator and destroyer, terribly strong and very weak, always existing and always disappearing – mean that it easily can be, and often has been, ascribed all sorts of different and conflicting symbolic meanings of fundamental importance at a number of shifting levels,[1] (d) the profound epistemological and ontological consequences of the fact that water is both nature and culture, since the thunderous liquid in a waterfall is the same water that is piped through cities; an inherent duality that highlights the importance of addressing how and why specific characteristics are attributed to different types of water, and underlines the fact that there is no mechanical, monocausal relationship between practical water experiences and religious water metaphors, and (e) to a greater extent than for other aspects of nature we can reconstruct long time-series of regular patterns and 'dramatic events' for water in ecological contexts because of water's ability to leave 'footprints' in the landscape, and because precipitation and river discharges have been of pivotal political and economic importance in the histories of most societies. In spite of the characteristics of water and its role in rituals and cosmologies, water has been given a peripheral place in research on religion.

A CRITIQUE OF TWO TYPES OF REDUCTIONISM
IN COMPARATIVE STUDIES OF RELIGIONS

Comparative studies of religions have quite a long history, but comparative and in-depth studies of water in religions have almost no history at all. This is, as this book shows, expected and natural, given the questions, concepts and analytical approaches that have dominated social sciences in general and this field in particular.

The historical and sociological study of religion has been heavily influenced by Max Weber and his *The Sociology of Religion* (Weber 1963 [1920]). Weber's influence was so strong because in an important sense his work represented a paradigm shift for the modern comparative study of religion. He analysed religion with other questions than the theological in mind and shaped what was later called the cultural-historical school. Weber's ambitious studies of religious traditions attempted to determine why certain cultures had evolved specific economic and social systems, and the role played by religion in that process. However, his impressive studies were reductionist in one important aspect: Weber proposed a research approach that was not interested in, or even disregarded, how religions were influenced by other 'situations' (his term) than those related to economics and the social, so ecology, and our experiences of ecologies or different waterscapes and water-society relations, were left outside of his empirical and analytical picture (Weber 1963 [1920]: 13).[2] The growing influence of Franz Boas and other anthropologists of culture and religion moved the focus of research further from the potential impact of geographical contexts and ecological experiences on religious texts and rituals (for example, Boas 1911 and Frazer 1922). Human cultures were regarded as self-contained, though interdependent, totalities, and in order to understand beliefs and rituals research should concentrate on revealing the workings of the human mind or minds that had produced the texts or rituals in question. The more theologically oriented traditions within the study of the history of religion have for obvious reasons not been particularly interested in how mundane, practical issues such as ecological 'situations' or adaptations have influenced creation myths, the images of Gods or formative ideas about heavenly power.

The result has been that while the phenomenology of religion established types, patterns and morphologies, these were not understood as being in any substantive way influenced by the physical context in which religions developed or operated, or by how people conceived and experienced them. This way of thinking has led to research designs that basically have been uninterested in such questions in general and in specific water-society relations in particular. The widespread priority given to texts over popular rituals has tended to overlook the pious enthusiasm for water and that rituals of 'the folk' all over the world have attached to religious acts and festivals in which water plays central roles (the water festivals in Asia in connection with New Year celebrations, the *Songkran* in Thailand, Epiphany in Christian-Orthodox countries, the *Rianovosti* in Russia, the *Makar Sankranti* in India, the *Pesach* in Judaism, dragon boat racing in China, and many, many more examples). The analytical approach proposed here, to study comparatively water systems and water-society relations and how they have evolved and been changed over time, does not restrict itself to those 'cultural' or psychological 'situations' on which Weber focused, but opens up the intepretative universe; it includes ecological contexts, situations and practices as well. By urging systematic comparisons of the views and practices of individual religions regarding the relationship between water, God and human beings we may also come to understand other similarities and differences between religions. By comparing these ideas and practices with the water-society relations and systems in which they developed, we might also obtain a better understanding of the complex interconnectedness between natural contexts and religious ideas in general.

I will in this chapter argue for the usefulness of the water-system approach in comparative research on religion, religious texts and religious practices in general. The proposal does not suggest reducing religious sentiment to impressions of admiration and wonder for water or claiming that water is or has to be an essential element in the conceptions of the divine. Sacred ideas should be distinguished from profane ones because they are of greater intensity, but also because they have qualities which other types of ideas do not have. The point here is trying to make sense of an empirical fact: most religions, but not all, give water a central but different place in the texts and rituals, in the past and today. Why is this so, and how can this be studied and what can such studies tell about religion in general?

Comparative studies of 'water in religions' may also help to liberate research from a certain normative hesitancy related to whether comparisons of belief systems are legitimate. Since water in most religions seems to be conceived of in more or less the same way, the idea that each religion is an organic whole with its own inner coherence, solely culturally determined by particular traditions, and therefore not comparable with others, must be qualified. With water as an entry point one might argue exactly the opposite – the both apparent and real similarities and differences in how religions conceive of water make comparative research useful and possible. One might extrapolate and focus on notions or beliefs about water because such notions are so common. Since water is such a widespread medium of myths and symbols, it is also easier to omit what has been described as a common problem in religious studies; that of applying one's own criteria of logicality and intelligibility to other belief systems and their corresponding criteria. We do not have to translate what is unfamiliar into what is familiar, since the different religions' orbits meet here, at the confluence of water, society and religion. A focus on the mundane water issue might therefore further a plural, cross-cultural approach to the study of religion. Water can function as a 'neutral', common ground, stimulating research on other and more contested areas. The study of ancient religions has long since been dominated by textual scholarship, which has given priority to the different text traditions,[3] but comparative studies of water in both rituals and texts might bring forth not only supplementary evidence but different perspectives.

CREATION MYTHS, GODS AND THE ROLE OF WATER

A better understanding of the creation myths requires research that breaks out of that kind of reductionism that looks only at social variables. Why was life according to the creation myths of Judaism, Christianity and Islam made possible when water mixed with clay, while in China life became possible when water was removed from the clay? And why, in pagan Norse religion, were there few, if any, ideas about the role of water in the creation of the world, and why was the Mayan religion's emphasis on water different from all of them?

Abrahamic religions share the basic ideas about water and God – God created the world and Man from water; God punished Man by water; and God's Paradise was a place defined by enough running water. The Old Testament and the Qur'an contain many passages in which fresh water is described as a gift from God and as a means of punishment. The Bible does not speak explicitly about the water of life but of God's river which waters the earth and creates nourishment and well-being: 'Thou visiteth the earth and waterest it, thou greatly enrichest it; the river of God is full of water; thou providest their grain, for so thou hast prepared it' (Psalms 65: 9). God, or Yahweh, is described as a fountain of living waters (Jeremiah 2: 13), and his blessings are compared in a variety of ways with the blessings of water: 'He leads me beside still waters; he restores my soul' (Psalms 23: 2–3). And: 'thou givest them drink from the river of thy delights. For with thee is the fountain of life; in thy light do we see light' (Psalms 36: 8–9). The opening incident in the Bible is man's loss of the tree and the water of life. The closing incident in the Bible is his regaining of the tree and the water of life (Frye and Macpherson 2004: 36).

Allah is described in much the same terms, and even 'His Throne was upon the Waters – that He might try you, which of you is best in conduct.' (Sura 11: 7) The Qur'an asks why people refuse to listen to Allah: 'And do they not see that We do drive Rain to parched soil, and produce therewith crops, providing food for their cattle and themselves? Have they not vision?' (Sura 32: 27). And moreover: 'It is God Who sends the Winds, and they raise the Clouds: then does He spread them in the sky as He wills, and break them into fragments, until thou seest Rain-drops issue from the midst thereof: then when He has made them reach such of His servants as He wills, behold, they do rejoice!' (Sura 30: 48). The name of the Islamic law, Sharia, means literally 'the path that leads to the watering place', that is, Sharia is the source of life; just as the watering places solve the practical problems of the Bedouin, Islamic law solves the problems of life and society.

In the Qur'an, metaphors about water are used to symbolise Paradise, righteousness and God's mercy. From the numerous references to cooling rivers, fresh rain and fountains of flavoured drinking water in Paradise, it is clear that water is the essence of the gardens of Paradise. The believers will be rewarded by 'rivers of unstagnant water; and rivers of milk unchanging in taste, and rivers of wine, delicious to the drinkers, and rivers of honey purified' (Qur'an 47: 15). The water in Paradise is never stagnant; it flows and rushes: 'In the garden is no idle talk; there is a gushing fountain' (Qur'an 88: 11–12).

Canonical, religious texts from many cultural areas underline the centrality of water in religious world-views and rituals. The Sanskrit text *Mahāābh ārata* (12.198: 14–19) summarises its general position: 'The creator first produced water for the maintenance of life among human beings. The water enriches life and its absence destroys all creatures and plant-life.' In the Puranic theory of creation, the Svayambhu (self-born creator) created water first. The old texts stated that primordial man was lying down in the waters of the universe (Sharma and Kanna 2013). In the book of Genesis it is said: 'In the day that the Lord God made the earth and the heavens, when no plant of the field was yet in the earth and no herb of the field had yet sprung up – for the Lord God had not caused it to rain upon the earth, and there was no man to till the ground; but a mist went up from the earth and watered the whole face of the ground – then the Lord God formed man of dust from the ground, and breathed into his nostrils the breath of life; and man became a living being.' (Genesis 2: 4–7). It was water that created the Garden of Eden, and it was the rivers running out of Eden that created the world for mankind. The Islamic story of the Creation has much in common with that of the Old Testament, and water permeates many aspects of Islam. The Qur'an states: 'We are made from water every living thing' (Qur'an 21: 31), and 'And Allah has created every animal from water: Of them there are some that creep on their bellies; some that walk on two legs; and some that walk on four' (Qur'an 23: 45), and he has ordained that all his created organisms will depend on water for life (*Qur'an Ayats* 24: 25). Although Yahweh, God and Allah created life by water, punished the human race by water in the form of devastating floods or droughts, and rewarded the believers with water in the afterlife in the form of a Paradise full of running streams and green watered pastures, and although ideas about water are central in creation stories and in narratives about 'the end of the world' in almost all known religions, there are surprisingly few comparative studies on water in religion.

Water seen as God's medium allows devotees to express and explain numerous and often incommensurable concepts of the world and the cosmos, and this cannot be explained, I will suggest, without studying and acknowledging the waters' varying physical capacities. The cultural history of the world has an immense pantheon of gods associated with water, and this must reflect not only the fact that water is universal in societies but also that it always manifests itself differently. Religious rituals involving water are also countless, and water rituals have been intricately interwoven with religious practices and profane activities throughout history. In all major world religions water is used to remove evil, to purge sins, to protect against future misfortunes and to enliven the spiritual dearth of everyday life. In many societies (but importantly for comparative studies, not in all), water has been seen as a force that cleanses the sins of devotees, be they Hindu pilgrims bathing in sacred rivers, Christians being baptised or Muslims performing their daily ablutions. The Qur'an describes ritual cleansing, the *faraid al-wudu*, in this way: 'When you come to fulfil the prayers, wash your faces and your hands as far as the elbows, and rub your head and your feet up to the ankles' (Qur'an 5: 6). Performing such rituals generally presupposes a certain degree of impurity in the practitioner, which must be overcome before or during ritual procedures, and purification with water as a neutralising force is what is needed. In the Bible, cleansing is very important: 'They shall wash their hands and their feet, lest they die: it shall be a statute for ever to them, even to him [Aaron] and his descendants throughout their generations' (Exodus 30: 21). Rituals may differ in form but the essence of the use of water is fundamentally the same: it is seen as carrying away both physical and symbolic impurity related to sin and defilement, and to the erasing of sin and the preparation for life after death (e.g., Parry 1985, 1994; Douglas 1994; Hertz 1996; Lehtonen 1999; Oestigaard 2005).

It is thus an undeniable fact that the physical, watery environment is often conceived of as a holy and cosmological landscape invested with divine meanings, where the profane and economic spheres are interwoven with the sacred. Rivers or bodies of water, for example, often have the role of marking the end of the profane and the start of a divine journey. Since the time of Pharaonic Egypt, it has been a common conception in many religions that on those who were immersed in water were bestowed divine qualities and grace. Also in ancient Indian religions dowsing oneself with water was a purifying action, while in Sri Lankan Buddhism, merely to look at water was sometimes considered to be cleansing. In many religions, bathing symbolises rebirth; it is a method of renouncing one's former self, but in other religions bathing has no religious value.

Water is also in general the medium whereby gods or God prove themselves or reveals that they are the god that they claim to be. The centrality of the rain gods in the religions of most traditions testifies to this fact. An impotent or powerless god will not be obeyed and worshipped, even if he or she is strictly speaking still a divinity, and the power of the gods is often measured through their ability to provide humans with life-giving waters in the form of rivers and rain (McKittrick 2006). An early and striking testimony of water's ability to prove the power and legitimacy of divinities is recorded in the Old Testament, where the cosmic drama and battle between the Jews and the worshippers of Baal unfolded on Mount Carmel (1 Kings 18: 16–45; Tvedt 1997: 85). Jahve proved to be the God who could control the water, a very important reason to choose him rather than Baal. Although the gods may exist ontologically regardless of their interaction with humans on earth, devotees have often perceived it to be the other way around. Water is also regarded as the primary materialisation of Vishnu's *mâyâ* (energy), and as a clear manifestation of the divine essence (Sharma and Kanna 2013), but very different from how Jahve manifested himself on Mount Carmel.

The procurement and control of water have to a much larger and more fundamental extent than the control of other aspects of nature been regarded as a divine project. In many religions the cosmos itself is created from water, at the same time as its role is described in different ways. In rainmaking rituals, this relationship between gods and humans takes a slightly different form. If the seasonal rain does not come when it should, the gods are invoked in the modification of nature for the creation of life-giving waters. Rainmaking rituals are rites where humans sacrifice to the gods for the return of water for a successful harvest and further life. In the *Bhagavad Gita*, for example, Mother Earth is a servant of God, and she is pleased when God is being worshipped. Rain, which produces all living things, is a result of the performance of ritual duties as taught or prescribed in the Vedic scriptures (*Bhagavad Gita* 3.14–15). The supreme powers of the gods are expressed by their divine control of water, which guarantees people's well-being and governs their life and their death by its presence or absence. A comparative study of water control as a divine project has, in spite of its importance, not yet been undertaken.

The scope of comparative research on people's relation to water and religion is so wide since water in religion symbolises or expresses the whole of human life in its various stages, but in different ways. Such studies should take as a starting point the fact that in some cases water in its original form is procreative: everything has its origin in and stems from water. Metaphors of creation and cosmogony have often obtained their strength and rationale in aquatic symbols where water has been experienced or conceived of as a procreative force and the essence of all kinds of life. In Christian baptism the initiate dies in the water by immersion and arises from it as reborn in the kingdom of God, while the precise role and description of water varies in different denominations. Rivers are often important symbols in religions, but not always, and again – in different ways. They symbolise the crossing-point between the living and the dead in the Pharaonic and Greek religions, but not in Christianity and Islam, In Hinduism, meanwhile, the river provides the mythical path leading to Nirvana, which is why the ashes of the dead should be scattered in a holy river. Running waters are often imbued with certain powers and qualities in the form of a spiritual or physical substance (Marriott and Inden 1974, 1977). In Hinduism, Ganges or Ganga is the Mother Goddess, and as such the water with its life-giving capacity is perceived and worshipped as a divinity (Darian 1978; Eck 1983; Feldhaus 1995; Oestigaard 2005). Ganga is the ideal holy river because she is the supreme goddess who may be used for every purpose; she is not only associated with the divine, but *is* the divine; she is not only worthy of spiritual respect, she *is* spiritual. There is no river like Ganga in Christianity, Islam or Judaism, although the Jordan was considered a holy river but in very different ways.

When comparing water's role in rituals it is important to consider aspects like the following: in Christianity, the water employed in baptism is not perceived as a divinity, but as consecrated water (Beasley-Murray 1962; Harper 1970). Although God transfers spiritual and divine qualities into this water through consecration by priests, the sacred powers are limited and defined for a certain purpose and time. Both types of water are within the realm of the holy, but their qualities and internal capacities differ. Ontologically, there is a fundamental difference as to whether the river *is* a divinity, as with Ganga, or whether the divinity transfers healing or blessing power *to* the water, as with the waters in the grotto in Lourdes. In Judaism, the 'living waters' do not represent an embodiment of Yahweh, but they do have spiritual qualities that allow humans to come closer to God. In Islam, the water of the Zamzam spring is Allah's own water; he made the water run in the middle of the desert by sending the angel Jibreel (Gabriel) there.

Water may be used as a point of entry for the clarification of differences between the holy and the sacred, and the divine and the sacred in new ways, since water is used for so many different purposes and in so many ritual connections. Even more so because despite all these different qualities in divine revelations and manifestations through water, in structural terms there are certain concepts that seem to recur in the beliefs and rituals associated with this element in nature and society. How are we to explain the importance of such similarities and differences, and how can we move beyond the isolation of certain elements of similarity to explore the deeper meaning and contexts of these similarities? The functional roles and forms taken by water in rituals have changed, and its use and how it has been conceived of therefore need to be analysed from a historical perspective.

In spite of a growing interest in aspects of water and religion, there are still relatively few scholarly works that attempt to provide analytical and general description of the role played by water in different religions, or of how water has been conceptualised and perceived at different times in different religions. Some studies have offered useful summaries of religious texts and quotations dealing with water,[4] but so far, none have dealt with the overall role and understanding of water in the different religions,[5] integrating analyses of texts, rituals and historical changes in the role and understanding of water in belief systems and religious practices. Although interesting studies have been published about aspects of water in different religions (Oestigaard 2013; Faruqui, Biswas and Bino 2001; Blair and Bloom 2009), we still lack comprehensive studies of 'Water in Christianity' or 'Water in Islam' or 'Water in Buddhism' or 'Water in Taoism' that integrate such textual and ritual analyses within a long historical and broad geographical perspective. There are studies of individual water rituals as in Lourdes and in Benares in India, but the bathing of Hindus in the Ganga or Christian baptism in water, or the fact that millions of Muslims bring water back home in plastic bottles from the Zamzam Well in Mecca every year, cannot be analysed by studying the history or the functional roles of these rituals in isolation, but must be related to textual analysis and differences in time and space between the waterscapes and water traditions within which the believers have lived. In a globalised world there is an even greater need – in order to provide a common ground for communication – for studies that systematically compare different religions, attempting to explain similarities and differences among ritual practices and textual narratives of core views. What are the preconditions for the co-existence of various concepts of holy or sacred water, of different water rituals, and of different conceptions of the role of water in the creation of the world? Water as an entry point provides a rare opportunity to study such symbolisms universally as components of religion and mythology, but at the same time within the confines of each individual religion.

Within the Jewish-Christian and Islamic traditions the notion of God's control of rain plays a central role. Rain can in fact be understood as the material symbol of the covenant with God. So long as the Israelites heeded the law, they received rain in reward. Or as it is written: '"though thou wast angry with me, thy anger turned away, and thou didst comfort me. Behold, God is my salvation; I will trust, and will not be afraid; for the Lord God is my strength and my song, and he has become my salvation." With joy you will draw water from the wells of salvation' (Isaiah 12: 3). The belief was that rain came from a great reservoir of water in the sky. It was God who controlled its release. Drought was therefore interpreted as punishment. By confessing one's sins one could placate Yahweh. It was only a placated God who would guarantee enough water for animals, wells, agriculture and extensive cleansing rituals.

The religion of the Vikings that for centuries dominated the belief system in the north-western part of Europe, gave water, however, a very different place in its cosmology. The Vikings' ideas about the creation of the world, about the origins of mankind, of paradise and the power of the gods were complex and fascinating (Steinsland 2005), but had no links to ideas about the holiness of water.[6] Here will be given a short description of their cosmology to show how different from the world religions it was in its conception of the place of water in the scheme of things, implicitly suggesting this should be interpreted as representing a mythical and religious reflection of the water-society relations of the Eurasian raincoast.

From the *Voluspá*, or the 'Prophecy of the Seeress', which was composed around the end of the heathen period, and the 'Gylfaginning' ('The Deluding of Gylfi'), which is the first part of Snorre Sturlasson's *Edda*, written in the thirteenth century, and paraphrases the older stories, and a number of other sources, one can derive an account of events as follows: in the beginning there was neither earth nor heaven. There was nothing except the great void, called Ginnungagap. This lay between two areas. One was freezing cold and foggy, and was called Niflheim. From Niflheim a river flowed into the void, where it froze into layer upon layer of ice. The other area was red-hot and was known as Muspelheim. At the point where the frost and the heat met there came into existence the first of the giants, called Ymir, and together with him a cow called Audhumla. While Ymir slept, his legs copulated with one another, and begat a son who became the ancestor of all the other giants of the earth. Meanwhile, Audhumla licked the salt off a stone. From this there sprang a human figure, Buri, who sired a son named Bort, who in turn sired sons who were called Odin, Vile and Ve. These three killed Ymir and created from his body the earth and the heavens. His bones became cliffs, his skull the sky, his blood the sea, and so on. Sparks from Muspelheim gave rise to the sun, the moon and the stars. The gods created the first man and woman, the first human beings, from some wooden sticks which they had found.

Ragnarok, or the twilight of the gods and the end of the world, would happen when the world was consumed by fire.

The south – that is, where fire and warmth came from – was associated with life. From the north came the rivers. These symbolised ice and lack of life. Yggdrasil – an ash tree – was at the centre of the cosmic system. In the Nordic creation myth the dramatic moment occurred with the meeting of fire and ice. Ymir was not created from precipitation or rain; life did not arise from flowing water, but at the point where heat met frost. Mankind was not moulded from the earth to which a god had added water, but was created instead from two wooden sticks. Paradise is not described as an area drenched in water. In Valhalla, where Odin gathered his chosen companions, the more important thing was mead. The end of the world does not arrive in the form of a deluge, as it does in Buddhist, Sumerian and Christian conceptions, but as fire and with the destruction of a tree. The lack of a flood myth and the marginality of water stand out as two of the most significant features of the Norse cosmogony, a feature that has been largely overlooked in research because this aspect of the belief system has not been systematically compared. In Scandinavian mythology, water as such had no holiness attached to it, and it was not a medium of the gods. It was a substance that could hide wisdom and spirits, but it was not itself spiritual.[7]

The Mayan religion in Meso-America should also be briefly discussed, since it has its own peculiar relationship with water. Mayan evocations of water deities are numerous and are always present in their iconography, their temple architecture, as well as in their rituals and written history. Water was one of the governing forces, as well as being the main sustaining structure of the world (Florescano and Velazquez 2002; Ruiz and Licea 2010). The divine condition of this element and the fact that it was understood as a symbol became a powerful way of understanding and cognitively expressing the world. Water became a central means of communication for Mayan communities, since gods and men could understand each other and come to sanctified agreements thanks to its divine essence. Their survival depended on this mystical dialogue about water. The God of Water proper, the giver of rain, was Chac, whose image is a human form with a huge hooked nose. The Mayas prayed to the god for the rain to be beneficial and to fertilise their harvests.

I will suggest that this religious belief system reflects the fact that Mayan civilisation was a rain-based agricultural civilisation, which stored rainwater in man-made reservoirs from one season to another. The Mayan heartland was a seasonal desert. The rulers of the Mayas were rulers whose legitimacy therefore largely depended on their god-like ability to maintain this water storage system and bring water or irrigation water to the farms. When the rains eventually disappeared in successive droughts during the eighth century, the economy was devastated. But it also impacted fundamentally on the whole political-religious fabric and authority structure of the society, since it was the leaders who should be blamed. In many historical studies the disappearance of rain in the seasonal desert of the Mayan heartland has tended to be overlooked, because the search for factors that can explain the downfall has been restricted to social variables.

China presents also a particular case of the universal society-water nexus in religion. China is known for having no real dominant creation myth; it was, as Joseph Needham put it, rather 'an ordered harmony of wills without an ordainer'. But basically, human beings lived in an anthropocentric universe where the sages brought order out of an originally chaotic universe. The water world was controlled on a grand scale, although the Jade Emperor, the mythical Yu, according to Mencius, guided the water by imposing nothing on it that was against its natural tendency. In China – whose religious tradition is marked by a syncretic blend of Taoism, Buddhism and Confucianism – one of the most famous creation or flood myths deals with Emperor Yu. It is connected to the Xia Dynasty of the third millennium BC. It describes a cosmic battle between flooding waters and the sky, the later conceived of as a dome, separate from the earth. One day, water emerges from the land and begins to rise up towards the sky. Two figures appear, a father and a son. They attempt to stop the rising water and restore the land. Both are described as being fish-like or dragon-like. The father fails and so the son, called The Great Yu, works for nine years to control the water and to dig channels where the water can flow. After titanic efforts to control the waters the land could re-emerge and society could be built. Most modern interpreters of this myth will suggest that this is the archetypical description of the flooding of the Yellow River. Water is described as a kind of primeval, mysterious force that needs to be controlled for the sake of the living. The semi-human figures that teach humanity how to control it are themselves watery, fishy or dragon-like in appearance, yet fully human. In religious studies this myth is often categorised as a 'flood myth', but this labelling should rather be interpreted as a reflection of the scholarly influence of the Abrahamic tradition in establishing the most used analytical and mythical categories, also conceiving the Jade Emperor and the creation of China and the world in this perspective.

To what extent should the clear differences among these religions in the roles they ascribed to water be regarded as a reflection of different spatial experiences with this water? The descriptions of the role of water that we find in the Bible and the Qur'an clearly correspond to beliefs that were widespread in the first great river civilisations, and that developed in the hot, arid regions of the Middle East. Illustrations and stories that survive from the times of Pharaonic Egypt tell how the priests already at that time were washing themselves before participating in ritual actions. Moses, whose name in Hebraic means 'he who came out of the river', and his people wandered around in the desert for 40 years according to the Bible, all the time dependent upon God's will to give them water, and Abraham's clan came from the valley of the Euphrates and Tigris, where both the Sumerian creation myths and flood stories focused on water. To what extent is the marginal role played by water in the creation story, in the end-of-time myths and in rituals in the Viking religions a reflection of the unique waterscape in north-western Europe, and of the fact that it was the religion that developed on the Eurasian raincoast? In Scandinavia and in Iceland there was more than enough water; in fact the problem was in general that there was too much of it, and the problem for the farmers was drainage rather than bringing water from rivers to desert sand as it rained all year round, if it was not actually snowing. In this context, water was conceived of as being less precious, and to dream of a Paradise of running water made no sense since in their earthly life the people were surrounded by running water day in and day out. And similarly, did the myth of the Emperor Yu gain its position precisely because the story reflected so well the experiences of the people on the Chinese plains, who had to adapt themselves to the recurrent, violent floods of the great rivers that now and then destroyed and drowned habitable land.

The point here is, of course, not to assert that there is a one-to-one mechanical and causal relationship between the ecology of the waterscape and the role of water in different religions. The world-views were developed in continuous interactions with the waterscape as part of a vivid and long-standing relationship. How much and to what extent such variables influenced belief systems and rituals is a task for future research to decide.

THE FLOOD MYTH – UNIVERSAL DREAMS OF URINATION OR REAL FLOOD(S)?

The myth of the Flood is probably the most studied of all myths. In the 1950s it was estimated that around 80,000 works in 72 languages had been written about Noah and the Ark alone. This astonishing level of interest is a reflection of the central place taken by the idea that God punished humankind with floods in Judaism, Christianity, Islam and many other traditional religions (Allen 1963; Leach 1969; Dundes 1988; Kramer and Maier 1989; Cohn 1996; Doniger 1998 and 2010).

The most dominant interpretation in the social sciences of these myths has been psychological, totally disconnected to any reflections on waterscapes or water-society relations. Comparative research inspired by a Freudian approach has been particularly interested in the dream aspect of the flood myths. Flood myths were according to this analytical approach products of the psyche which emanated from a universal trait of the human soul. It was suggested that there was a connection between dream responses to the basic need to urinate during the night and the ubiquity of the flood myth. This perspective produced many scholarly articles and described the spread of the flood myth as a kind of retelling of such disturbing dreams. Others have given the myth quite another psychological explanation; they see it as a male chauvinistic or patriarchal dream. A masculine god rescues the world and makes a pact with a male survivor, or masculine hero. The claim is that it is a creation myth, modelled on, or formulated in response to, the 'female flood', that is, the 'water that flows' in connection with birth. Just as mankind is born of woman, so the world is created, or born of, man. It has also been seen as a metaphor – 'a cosmogonic projection of salient details of human birth insofar as every infant is delivered from a "flood" of amniotic fluid' (Dundes 1988: 1). These psychological and generalised interpretations of the flood myth take for granted that the myth and its story have been diffused (Dundes 1988: 2), and ignore the differences in the position that the myth has occupied in various religions, and in the character of the different doomsday conceptions in different mythologies. To advance our understanding of the flood myths it will therefore be fruitful to carry out further and more rigorous comparative historical studies of water-society relations and how they have evolved and been reflected upon.

The relationship between the much older Sumerian flood myth and the myths of the Bible and the Qur'an is now beyond dispute. Archaeologists have found evidence not just of one, but of many floods in the region, and it has been established that the deluges that affected places such as Ur, Kish and Uruk cannot all be dated to one and the same period. Most researchers believe that some of these floods resulted in serious destruction and made such a deep impression that they became an enduring theme in cuneiform literature. In the course of time, these different stories were transformed into the single story of the Great Flood. The prophets of the Middle Eastern monotheistic religions regarded the thought of an angry God who wanted to punish sinful mankind by cleansing the world and making a new start as an eschatological inundation. The waters sent by God would cleanse both land and people, wipe away faithlessness and plant a new spirit in the hearts of mankind. Water was duplicitous: both life-giving and threatening. It was the medium through which the gods could distribute blessings and punishments. 'I will bring a flood of waters upon the earth, to destroy all flesh in which is the breath of life.' (Genesis 6.17)

Other civilisations also have flood myths whose narratives are reminiscent of that of the Bible. Around 300 comparable stories have been counted worldwide. The Hindu flood myth, although not associated with God's punishment, is reminiscent of the Jewish and Sumerian versions. The Lord of all Creation, Brahma, revealed himself in the shape of a fish to Manu, the first human of Indian mythology (as part of an Indo-European language, 'Manu' is related to our word 'man'), and told him of the coming flood that would destroy all things. He advised him to build a ship, and in the hour of danger to go on board, taking with him corn which could be sown in the earth. Manu did as the god advised and harnessed the ship to the fish. Guided by the god, he eventually landed on the highest peak of the Himawan Mountains, where, in accordance with the god's promise, he came to rest. When the flood receded, Manu offered sour milk and butter to the waters. A year later, a woman was born who was called 'Manu's daughter'. Together the two of them rescued the human race. In Hinduism as in Buddhism there is no ultimate destruction or dissolution. It is a continuous cycle of creation, dissolution and recreation from the dissolved condition. The whole cycle in these religions resembles of course the seasonal pattern of birth and destruction that has been so characteristic of 'Monsoon Asia' and where the floods have tended to be very destructive, setting land under water for weeks and months on end, but at the same time being necessary as the beginning of the next growing season. One explanation for these differences must be sought in two aspects of the physical waterscape and their relevance to societies: the floods were a regular, annual phenomenon in Monsoon Asia, sometimes being very destructive but with everybody knowing that things would revert to normal 'next year'. The Sumerian cities had developed not only by adapting to the natural variability of the rivers' water (as for different hydrological reasons was the rule in the Hindu cultural area) but by controlling it and even channelling it. A great flood was therefore much more destructive in Sumer, attacking so to speak the very heart of the society's achievements and economy, and it was therefore more logical to intepret water's destruction as punishment of the people by an angry god.

In Norse mythology there is no flood myth like those which are found in the Bible, the Qur'an or the Epic of Gilgamesh. Forty days and 40 nights of continuous rain were fairly normal for the people living on the Eurasian raincoast also in the time of Odin and Thor, so torrential rainfall in 40 days could not be interpreted as the end of the world. 'Ragnarok', the Norse 'end-of-the-world' story, was preceded by three terrible cold winters and the Sun, fighting a desperate struggle, was eaten by the wolf Fenrir. The Japanese Shinto religion had neither the concept of the world coming to an end nor an idea of a global disaster in the form of a great flood. In the Pharaonic religion of Egypt water played a very central role, but there was no story of a deluge that destroyed everything.

Some researchers (and creationists) have been looking for a general world-wide inundation caused by rising sea levels as the explanation to the centrality of the flood myths – that there was one global flood event as the background to them all. It is argued that the assumed consistency among flood legends found in distant parts of the globe indicates that they were derived from the same origin. Others have promoted the hypothesis that flood stories were inspired by a kind of observation of seashells and fish fossils in inland and mountain areas (Mayor 2011). But as has been indicated, neither the Egyptians nor the Vikings had a flood myth, and the flood myth of the Hindus was very different from the myth of the Gilgamesh Epic. There are more than 500 myths known to us that portray a flood in some way, and they do it in highly diverse ways. It is more natural and logical to see these as stories told about real floods that happened in the past along different river basins, often dramatically affecting the lives of people who had settled on the riverbanks.

The recent trend of looking for changes in sea level as the background to these myths is not very fruitful but speculative. Authors have started to discuss whether Plato's story about Atlantis actually happened (Castleden 1998). Some have suggested that the story might reflect that the geography of old Mesopotamia was considerably changed after the last Ice Age when the sea level rose and filled the Persian Gulf with water. Another hypothesis is that the meteor or comet, which supposedly crashed into the Indian Ocean around 3000–2800 BC, created a giant tsunami. There has also been speculation about a devastating tsunami in the Mediterranean Sea, caused by the Thera eruption, but research has indicated that this had a local rather than a regionwide effect. It has been postulated that the deluge myth in North America may be based on a sudden rise in sea level caused by the rapid draining of prehistoric Lake Agassiz at the end of the last Ice Age, and one of the latest hypotheses about long-term flooding is the Black Sea deluge theory, which argues that a catastrophic deluge happened about 5600 BC when the Mediterranean Sea flooded into the Black Sea. Many of these events may have happened, but these localised floodings cannot explain the actual distribution of the flood story and most likely the chronicles of the first civilisations would have mentioned such extremely dramatic events.

In the era of emerging agricultural civilisations in dry valleys dominated by violent rivers, it is more natural to look for the actual ecological background to such stories in the imbalances of water-society relations at the time. Many of the excavated cities of classical Mesopotamia, where the legendary walls of Uruk and Shurrupak were created on the banks of the Euphrates, present evidence of flooding, but at different times. Archaeologists have been searching for evidence of such a flood in Israel (Bandstra 2009: 59–62), but there is of course no such evidence of a widespread flood, because this area of the world did not have a flood-prone waterscape. No story of a deluge existed in Pharaonic Egypt, while there was most likely one in the Greek and Roman period, but the papyrus that contains it is damaged and unclear (Frankfort 1948; Budge 1989 [1923]: ccii).

If metaphors in religious texts are not to be seen simply as an ornament of language or as a controlling mode of thought expressing psychological mechanisms, then the flood myths can be interpreted as reflections of social experiences. In this case, the distribution and character of the waterscapes and the water-society relations must be part of the interpretation. The thesis would be that these flood myths emerged in countries with violent floods and marked differences between wet and dry months, but not only that: they were most important and punishing in areas where people lived along river systems and where they had developed the art of water control. Flood myths originally played no role in the apocalypse myths of people such as the Vikings, who inhabited regions of the world where rivers tended to run more or less all year round and where great floods were rare and never particularly serious and did not dramatically affect settlement patterns and economic activities. Neither did myths of a destructive deluge play any role in Pharaonic Egypt, where the yearly inundation was a blessing and they therefore had different flood myths, cultural-specific and reflecting the character of the regular and slow flooding of the river, nor did they in Japan, where floods were comparatively rare and modest in scope and destructive capacity.

Scholars have, of course, presented different theories about the relationship of flood myths to ecological experiences. Some have argued that the fact that so many tell the story of Noah must reflect some kind of societal considerations of experiences of an actual catastrophe happening on a global scale. Few, if any studies have on the other hand systematically analysed and compared the water-society contexts of the emergence of the different flood myths, integrating in the analaysis the ambivalence that water represented and symbolised for those who lived at that time and in those areas where the stories were first told and written down.[8]

My proposition is that in order to understand the flood myths and how they emerged and were diffused, and to shed new light on the relationship between geography or ecology on the one hand and myths and religious rituals on the other, comparative research on the character and relationship between 'end-of-the-world-stories', waterscapes and experiences with different waters would be fruitful. These myths should definitely not be treated simply as synonymous with the illusory. Their dual naure is based on a past reality and, pointing to deep experience, the

81

threatened destruction and the hope of renewal reflect both the character of an actual flood and the character of water in real life, but a character that is more prominent in some places than in others. This ambivalent power is what the theologian Rudolf Otto called the *mysterium tremendum et fascinans*, the water 'mysterium' that terrifies and fascinates and thus produces mythical stories. Do the myths then build on historical events or are they fictions? Was there really a global flood, as some scientists will argue, or are the story of Noah and all the other similar myths based on collective memories of real regional or local floods (Doniger 1998, 2010)? If there was a global flood, why then do not all religions have a flood myth of some sort? On the other hand, is the almost global occurrence of the myths due to their symbolic content rather than a shared experience, or are they widespread because floods are widespread?

The long traditions of comparative cultural and religious studies of the flood myths should be broadened and should integrate more historical data about hydrological conditions and existing man-made water modification structures. Based on compilations of historical data on climate, river discharge series, rainfall patterns as well as on water control measures and installations, no matter how rudimentary they are compared with modern achievements, the possibility of finding more definite answers to questions such as whether the different myths and doomsday stories were related to perceived history or experienced ecology, to fiction or to metaphor, would be greater. The flood stories and the natural and modified waterscapes and their roles in which they were told must also be analysed in wider textual contexts, since the drama of the stories and their meaning can only be properly understood as part and parcel of how the central relationship between divinities and water in general is described in the canonical texts of the religion concerned.

WATER AND THE RELIGIOUS 'BLAME GAME' OF ECOLOGICAL DISASTER

A focus on relationships between water and the divine can also make more general analyses of the ecological attitudes of world religions more precise and empirically rewarding.

When Lynn White Jr. published 'The historical roots of our ecological crisis' in *Science* in 1967, he initiated a very influential debate about religions and the ecological crisis. White argued that the Judaeo-Christian tradition must bear responsibility for this crisis, because of its dualistic view of Man and Nature, where Man stands above and apart from Nature, while men and nature in other world religions were part of the same web created by the Almighty. White ended up wanting to reform Christianity, making Francis of Assisi the patron saint of ecology, and consciously attempting to construct an alternative Christian environmental ethic. Comparative studies of water and religion could shed new light on this issue.

Since White's seminal article quite a few studies have been published on the views of religions on nature and on ecology. Typical titles have been 'Is it too late? A theology of ecology' (1972), 'Ecological problems and Western traditions', or 'Can the East help the West to value Nature?' (1987). The Harvard Institute of Social Action on Religion's programme on religion and ecology is especially interesting in this context. What these impressive studies have demonstrated is that there are methodological problems involved in trying to 'identify and evaluate the distinctive ecological attitudes, values, and practices of diverse religious traditions, making clear their links with intellectual, political, and other resources associated with these distinctive traditions' (Tucker and Grim 1993: xxi). Ecology and nature have been defined in extremely broad terms, covering almost everything.[9] Typical questions within these traditions are therefore posed awfully broadly, exploring, for instance, the ways in which 'different religious perceptions and cultural values affect human beings' understandings of their relationships with nature, and their actions in and upon the natural environment' (Arnold and Gold 2001: xiii), or 'How do human beings in different cultural worlds think through and about their relationships with the natural environment in which they live, work, eat, pray, give birth, die' (Arnold and Gold 2001: xiv). The problem is one that will be discussed more in depth elsewhere in this book; such concepts and terms as 'nature' and 'ecology' are extremely broad and carry contradictory and unclear connotations, and have, moreover, different meanings in different cultures and religions. Additionally, no religion has similar attitudes to all aspects of the surrounding nature or ecology precisely because 'nature' and environment' mean things like animals, stones, water, sun, wind, plants, humans, and so on.[10] Using these terms as the basis for comparison and analysis makes it possible to argue in favour of all kinds of general conclusions, because it is always possible to find examples that illustrate or strengthen one's own arguments. This empirical and conceptual problem is aggravated by the fact that the question is deeply affective and motivational.

A focus on the role of water in comparative studies on religion is much more manageable; it is researchable. It will also falsify White's thesis, since the role of water in different religions undermines the general thesis about Judaeo-Christian traditions and the way these understand nature. In Buddhist and Taoist China, for example, the dominant stories deal with the manipulation of water on a really grand scale, and much more than in Christianity and Judaism. As the Chinese sage Lao Tze said about 3,000 years ago: 'The wise man's transformation of the world arises from solving the problem of water.' The Hindu literature has many more examples of humans trying to influence the gods to change the water landscape and precipitation patterns in man's favour than Judaeo-Christian texts. In the Qur'an, water is God's water, just as it is in the Bible. Man does not stand further 'above' water in the Bible than he does in the Qur'an or in the Baghavad. To the extent that man is aiming at controlling the watery nature within the Judaeo-Christian tradition, any claim that it is more geared towards mastery, taming and control than was the case in the old Egyptian religion, the Chinese religion or Islam cannot be sustained. Since water is such an important aspect of all ecosystems and of societies' relations to the environment, theoretical arguments of fruitful relevance about world religions' attitudes to nature must also be evident in attitudes to water. The theory or claim that the Judaeo-Christian tradition has a more instrumental relation to nature and thus also to water, or to water and thus also to nature cannot be sustained. There is a need for much more systematic comparative research on the whole web of practices, water festivals and water rituals, and on how water is described in texts and reflected in iconography to be able to formulate a precise thesis on these very important issues. Concentrating on water, as a single aspect of nature on the basis of methodological arguments about what can be studied and compared, could make comparative research on religion and nature more rigorous and controllable.

THE 'SECULARISATION' OF DIVINE RUNNING WATER

Water-society relations will also be a fruitful entry point to a better understanding of how rituals are affected when the ecological contexts of believers change – an issue of growing importance in a world of increased global migration and technological developments. Here two cases are briefly discussed: Christian baptism from the River Jordan to the Norwegain raincoast close to the North Pole and holy rivers in Hinduism from the Ganges to an industrial river in England.

Christian baptism is still described as the bath of rebirth, although baths in rivers are seldom involved in mainstream Christian rituals today. In Christian baptism, water plays a key role as a symbol of renewal and resurrection. Baptism in water is described and understood as the action whereby God helps the individual over from the worldly realm to that of his own Kingdom, from the world of sin and into the community of Heaven. The New Testament specifies that the baptismal ceremony is to be carried out in the name of the Holy Trinity, and that water is the element which serves as the medium.

It was St Paul who institutionalised Christian baptism with water, based on the example of John the Baptist and his baptism of Christ in the River Jordan. It is to him and the period in which he lived that we must look in seeking the origins and background of this ritual. In Palestine and elsewhere around the Mediterranean, water was a scarce resource and was therefore generally a highly valued symbol of life and divine mercy as discussed above, and as reflected very clearly in the Mikwah, the Jewish tradition. But these climatic factors are not sufficient to explain the nature of the ritual. In St Paul's time, an extremely popular cult of great influence throughout the Mediterranean region was that of Isis–Serapis or the Nile cult. This religion, a version of a much older Egyptian cult of Isis and Osiris (Anthes 1959; MacQuitty 1976), spread during the first century AD through Asia Minor and into Greece, and when it reached Rome in the course of the second century was a competitor with Christianity also in terms of the number of adherents it attracted. The cult became so popular that on several occasions citizens of Rome forcefully resisted decisions of the Roman Senate to tear down its temples. The extent of the cult's influence on early Christianity is still a matter of debate. Some reject the idea that St Paul's precepts concerning baptism are adaptations from this cult, yet there seems to be a growing consensus that they at least are strongly connected. Christian baptism is thus influenced by the great significance of sacred water in the region in general (Nile water and, later, Jordan water), but also by the old Jewish tradition of the bath of conversion, that is, the ritual bath which non-Jews had to take when converting to Judaism. According to a number of historical sources, in the first century baptism was supposed to take place in 'running water' or in a river. The ordinary apostolic mode of baptism was immersion, clearly representing death and burial with Christ, followed by a resurrection to new life with the resurrected Christ (Harper 1970). The descent into water and the rising from it corresponded to death and resurrection.

The question whether immersion is a necessary part of the ritual has of course been a long- and hotly debated and conflictual issue within Christianity. All agree that the essential feature of the ritual was water, but there has been disagreement about the mode of its use. Some argue that the insistence upon form contradicts the Scripture and the temper of the age of John, Jesus and Paul. Those who have argued in favour of a focus on the essential role of water more than its form have made the point that the ritual must be adapted to ecological circumstances and local waterscapes (Lambert 1903: 225; McGiffert 1897: 542). When Christianity expanded into north-western Europe, where water did not have an aura of holiness and where it also became more difficult due to climatic conditions to perform the bath, baptism changed.

It became less and less frequent for the baptismal ceremony to be held outdoors, even if this was the principal practice up to and during the fourth century. The first documented case of a new mode – that of affusion (pouring water over the head) – was around AD 250 (Russell 2001: 25).[11] Eventually, the ceremony moved indoors, and became confined to the churches. As Christianity expanded into northern Europe, affusion became the usual manner of administering baptism. By the thirteenth century, wetting had taken over as standard practice throughout the Roman Catholic Church, although for a long time it still remained important to use 'running water'. In Latin-speaking countries, and in those influenced by Latin culture and language, the baptismal stoup was usually described as the *fons*, or the font (cf. fount and fountain), in other words, it remained associated with the running water of a spring. The sacred quality of water was at that time still associated with the idea of it being in motion.

The importance of this idea of running water as the most holy is also demonstrated by the evolution of the baptismal font in the history of church construction. Initially the font was of a size that allowed the child to be fully immersed three times in the water. In the Middle Ages fonts generally had a hole in their base, which allowed the water to run out through the pedestal, through the church floor and down into the earth. The hole was plugged before the bowl was filled with water. After the ceremony the water was released into the earth, for having served in baptism it was considered to be so full of divine power that it would have been sacrilegious simply to throw it out with the slops. As immersion was gradually replaced by affusion, fonts grew steadily smaller, and it is now a long time since fonts were built with their own drain pipes. Nowadays, the water – still described in the actual ritual by the priest as divine – is tap water from the nearby kitchen or bathroom. The water itself is not in general seen as divine any more (although there are exceptions to this rule), but the language about the water in the ritual is the same as it has always been.

The content and symbolism of Christian baptism have clear historical roots and were originally influenced by cultural, economic and social relations between people and water in the Middle East. And with the spread of baptism to parts of the world where water conditions and temperatures are very different from those in the Middle East, a situation arose in which the significance and role of water also changed. Today's rituals are a distant and much transformed reminder of these 'foundational' circumstances. The role of water in the rituals changed as the waterscapes changed and what remained acquired an increasingly symbolic content and meaning.

In Hinduism the River Ganges plays, as we all know, a crucial role, and the notion of holy rivers is central in a great many rituals (see, e.g.,

Darian 1978). Some research has been done on ritual adaptations to holy rivers when they become dangerously polluted. When, at the beginning of 2000, this happened to the Bagmati, the holiest river in Hindu Nepal, as it was running by Pashupatinath in the capital Kathmandu, the believers were told not to bathe in the river but instead to take the waters in showers erected on its banks instead.[12] But what happens with these rituals when Hindus move away from their traditional holy rivers to new countries where the landscapes obviously do not have the mythical dimension that is ascribed to the rivers of the Indian subcontinent in the Hindu texts? Will the rituals change and, if so, how, or will the rivers at the new places where Hindus live be given a religious character, and how will these practices be religiously sanctioned and justified in the short and long run?

Bradford in the UK provides an interesting case. The River Aire is a polluted, industrial waterway that sluices through Bradford. This is an unlikely spot in the Hindu cycle of reincarnation. But the local Hindu population sought permission from Bradford City Council to turn the river into a 'symbolic' Ganges: a Ganges substitute. The Ganges flows more than 2,000 kilometres from northern India to Bangladesh. The River Aire comes into life north of Skipton in the Pennine hills and runs a mere 160 kilometres before it empties into the River Ouse. The idea was that a Hindu priest should pour a little water collected from the Ganges into the River Aire, and then the Hindus could scatter their ashes in what was directly described as a substitute river.

The important issue in this context is not that the Bradford City Council did not concur with the plans and initiative. The question is: how could this ritual be justified and ritualised by the devotees in relation to the River Aire? Water has a particularly great potential as a religious medium, also because, unlike ordinary relics, it can very easily be used to transport and diffuse holiness from one place to another. Since there is always so much of it, nobody – neither church nor priests – can totally monopolise the control of this symbol of the sacred or of holiness. It is possible to lock up fragments of relics guarded by officialdom, but the fluidity of water usually evades such attempts at control. Holy water is and has always been more accessible to the general population, and this must be one reason why water rituals in many situations have become a kind of 'people's religion'. It has been possible to infuse new meanings to new rituals because the rituals themselves can be performed outside the control of the religious hierarchy (also after the introduction of Christianity to Europe, the tradition of holy wells and holy water persisted long into the nineteenth century in most places, including England and Scandinavia, in spite of the fact that the practice was forbidden). To what extent will the process that was foreseen in Bradford be similar to earlier developments in Asia, when Hinduism spread out from its birthplace and across the ocean to Indonesia, and how, for example, can the history of the establishment of Lake Manasarovar, far up in the highlands of Tibet – a very holy lake for Hindus, Buddhists and Jains – be reconstructed by studying Hindu texts, pilgrimage and the particular physical and social qualities of water?

What these examples show is that water myths and water rituals differ enormously from place to place in their morphological character but can still, at least partly, serve their social and religious functions. Water ideas and water rituals are not a 'closed' category with the same characteristics in different cultural areas or physical environments. There are a number of similar cases that have not yet been studied and that are therefore not yet properly understood.

MODERNITY AND HOLY WATER

The conventional and very powerful notion that nature idolatry is something belonging to the past or is gradually fading away in the wake of modernity is contradicted empirically by the role of water in contemporary society and belief systems. Never before have so many people taken part in religious rituals where the use of some form of holy water is at the centre of the rite. Millions and millions go to take holy water or holy baths, or to receive God by being baptised in water. Every year some 3 to 5 million people journey to Lourdes at the foot of the Pyrenees (Gordon 1996; Harris 1999). No other place in the Christian world, apart from Rome, receives so many pilgrims. They come from all over the world to this small French town with its holy spring and healing water. It became a place of pilgrimage after Bernadette, the young daughter of a local miller, saw the Virgin Mary creating a spring in the muddy soil. Every year millions of Muslims on the Hajj pilgrimage to Mecca go to the Zamzam Well, a water source miraculously generated by God. One story has it that God sent the Angel Gabriel who kicked the ground with his heel and the water emerged and Abraham's son was able to drink. Due to modern technology like plastic bottles and aircraft, believers can now easily take home the cherished water. India is the land of water pilgrims *par excellence*, not only in terms of tradition, but also because of the sheer scale involved. The most important festival is the Kumbh Mela, which is held every 12 years at the confluence of the Rivers Ganges and Yamuna and the mythical Saraswati. During the last Kumbh Mela 120 million people gathered over 55 days, the largest congregation of human beings that the world has ever experienced. This mass of people came to the same place with one purpose: to bathe in the confluence of the holy rivers. Within Christianity, Pentecostalism is the fastest-growing denomination, and one of its most central, distinguishing rituals is baptism with the Holy Spirit by immersion in water.

The sheer number of people who currently take part in rituals where water is at the very heart of them, makes comparative research on the religious meaning and role of water also highly relevant for understanding the religiosity of today's world. Few things reveal to a greater extent the notion that nature idolatry is something of the past, a modernist fallacy. This salient aspect of modernity must be explained by a combination of factors, but it must also take into consideration those qualities of water that have made it and still make it natural for humans to spin webs of significance around it in ways that no other element in nature can match. Taking water as a point of entry, let us study structural similarities and diverse empirical differences in religions in a rigorous, comparative way, which can contribute to making the study of religion a meeting ground of complementary methods. Comparisons of water systems and religions offer a unique opportunity for the integrated and comparative study of texts, rituals and practices, thus improving our understanding of the relationship between ecological contexts, religious ideas and dogma in general. Such research will also be of practical concern in countries that face serious challenges related to their water resources, assuming that religious beliefs and ideas about water have a bearing on attitudes to water management. Since many of the great civilisational and transboundary rivers are shared by believers of different religions, such as the Ganges, the Indus, the Donau, the Nile and the Mekong, the role of religious ideas about water is also a question of global and current, hard-nailed geo-political concern.

5

BETWEEN THE HYDROLOGICAL AND HYDROSOCIAL CYCLE: THE HISTORY OF CITIES

Some of the first and still some of the most famous descriptions of urbanisation underline the role of water: Herodotus' fifth-century BC descriptions of cities located on the banks of the long-gone branches of the Nile in the Egyptian delta retain an importance the author could never have envisaged: his notes of these cities long since drowned by history and changing relations in water-society relations have become invaluable sources for archaeologists and for climate scientists who are trying to reconstruct the development of the Mediterranean coastline.[1]

Somewhat later, but still 2,000 years ago, the Greek geographer Pausanus travelled throughout the Mediterranean region, and after visiting the urban centres of his day, wrote that no city could claim to be a real city without a fountain in its midst. Cities in this relatively dry part of Europe, where months could pass without a drop of rain falling, 'revolted' against their geographical situation by installing fountains. The fountain was the ultimate symbol of urban life; it marked the distinction between city and countryside and was a powerful, symbolic expression of the triumph of culture over nature. It was also a very visible sign of what has proved to be a structural and enduring aspect of the city as a social phenomenon: this type of social organisation requires quite complex control over water in one way or another.

This chapter will argue for the relevance of bringing water-urban systems into the centre of urban studies based on two basic premises.

First, all urban dwellers – from the few people who settled at a spring in the desert and built a wall in Jericho almost 10,000 years ago, to the Incas living in the royal city of Machu Picchu on its mountaintop in the Andes, with water carried by long aqueducts built along the mountainsides, to stock-traders living in apartments in a skyscraper in Manhattan – share the absolute need for *one* controllable resource: H_2O. The theoretical and empirical importance of the fact that all urban dwellers – rich and poor, white and black, Hindus and Muslims, Christians and atheists – need the same amount of water can hardly be overestimated. And, moreover, as long as people live in cities, water has to be socially provided, by canals, pipes, rainwater, aquifers, rivers or desalination plants. Its control, distribution and disposal have to be technically organised and socially managed. Water is the only absolutely essential and universal urban resource that can be controlled, and all urban settlements have therefore always struggled to do this, and in this effort, urban spaces have developed and changed and cities' power relations been cemented and dissolved.[2]

Secondly, it is just as crucial to draw the analytical and theoretical consequences of another physical and social fact: the hydraulic system that envelops the city varies from place to place and from time to time. Spatial differences in precipitation and evaporation patterns and in urban water landscapes fundamentally define the character of cities as well as the sense of place. While all cities need to solve the water issue, they have to do so in different ways, since the urban dwellers' interactions with, and their patterns of activity in relation to, their water will reflect the local water cycles' particular characteristics and the past interactions of water and city.

All cities have been and are locked into this continuous web of relationships with water's simultaneous universalism and particularism. The history and current development of cities is therefore written in water and in the most varied ways and manners. In spite of this, mainstream urban studies have persistently tended to neglect the water issue and the interactions between urban development and water and how these in fundamental ways have impacted on the whole process of urbanisation.

THE 'DRY' TRADITION OF URBAN STUDIES

Charles Darwin wrote that science must be criticism before it can be anything else. In line with this attitude this chapter will first discuss the dominant tradition within urban studies, a tradition that has been reductionistic, in the sense that it has mostly been concerned with certain social variables and it has tended to become deterministic when natural variables were included.[3] During the decades after the 1960s, when urban studies were established globally as a research field, it by and large completely neglected the importance of the confluences between the planet's water and cities and how these interactions had fundamentally impacted on urban life. It is here thought crucial to shortly discuss this research profile, in order to make it easier to open the door to less reductionist research strategies that can broaden our understanding of urban development.

It is generally acknowledged that early urban studies in the first decades of the twentieth century were concerned with 'site and location' issues and with nature's physical characteristics as a determining factor in the location and development of settlements. At the time, a number of descriptive studies of city and town establishments were published, emphasising the roles of topography, water, climate and other physical factors. The aim of much of this research was to demonstrate that the character of towns could be derived from their physical locations. However, this research tradition was later discarded not only because of its determinism, but on the broader grounds that it was concerned with the 'causal effect of physical geography' (Carter 1981: 3). Mainstream urban studies have since argued that it was the replacement of this type of causal description concerned with urban-nature relations with interpretation of the social that laid the foundations for modern urban studies. The 'site-situation' approach was criticised for being theoretically wrong, but importantly for later research practices, it was also regarded as historically outdated due to the development of the city itself. Nature had been overcome as a relevant factor in modern, large-scale, urban development, and to focus on it was therefore both theoretically and empirically anachronistic.[4] This fascination for the social aspect was rational, as cities expanded in all sorts of directions and transformed themselves into complex societies of their own.

Concern with physical factors was thus seen as both historically outdated and theoretically immature. 'Mature perspectives' were not interested in nature or in physical factors, and mature research should only deal with how social facts constrained social action or, as it has been formulated, with 'the way in which urban patterns and processes are the outcome of the combination of human choice and wider social processes which place constraints upon this human action' (Hall 1998: 20–1). Urban choice and constraint were seen as being entirely within the social world, unfettered or at least not significantly influenced by non-social factors as nature or waterscapes. In the post-1950 period, the clearly dominant subject matter of urban geography has been defined as the set of topics that can be explained in terms of social variables, and social variables alone. Within this theoretical perspective and conceptualisation of the research field, water systems, being physical phenomena of relevance to urban development, were simply of no interest. Also when water clearly functioned as a social variable, in fountains, in coffee shops, in urban drainage and in sewage systems, it was still categorised as nature, and was on this ground outside the empirical universe of the urban scholar.

In the first decades after 1950, urban studies in general were influenced by positivism and the aim was to detect universal social laws and fundamental regularities and to build models.[5] Proponents of what was regarded as a scientific method in human geography, for example, argued that 'geography has to be conceived of as the science concerned with the formulation of laws governing the spatial distribution of certain features on the surface of the earth' (Schaefer 1953: 227). These 'certain features' were all social features, and Schaefer and his colleagues criticised urban geographical enquiry up to that point for being analytically naïve and insufficiently interested in social theory. This criticism was of course to the point, but this nomothetic-oriented tradition, only concerned with the social and social variables, could on the other hand not take the extremely varying roles of the nature of water in city development aboard empirically, because to do so would erode its model-building ambitions.

Two theoretical schools that were uninterested in the role of nature (and water) came to dominate the field in this effort to establish a positivist science, seeking to construct theories of spatial laws on the basis of statistical analysis and the construction of predictive spatial models. The 'ecological approach' was based on the assumption that human behaviour was determined by what were described as 'ecological principles'. This term was chosen in order to emphasise the idea that urban developments should be analysed as analogical with the biological world, because the same behavioural rules that governed the biological world determined predictive patterns of urbanisation. The social world was seen as a mirror of the biological world, but the impact and role of nature or of water was of no interest for the study of urban development.[6] The 'urban ecology' literature was solely concerned with the social and with understanding the relationships among various competing social groups living in similar or separate areas, using physical ecology concepts.

The 'neo-classical' school was based on the assumption that human beings were motivated primarily by rational aims, and that their patterns of actions were therefore predictable. Actors were consequently locked into an eternal effort to minimise costs and maximise benefits. In line with this view, urban developments were studied as a function of this rationality, and nature or physical elements in nature were logically 'excluded' and of no interest to the analytical set-up.

A salient and shared aspect of both the 'ecological approach' and the 'neo-classical' style was that they were uninterested in water-urban systems and relations. They were solely concerned with the social and with social variables defined in sociological terms. This indifference to the water issue in general extended therefore to water also when it was part of the social sphere. Urban studies, whether we are talking about sociology, geography or history, did not at the time describe or analyse water in urban space as part of the social world – be it as an object of worship, as a mediator of social power or as an area of technological discipline and development – or how changing notions of water reflected social and economic interests and affected urban health and urban health management, and the architecture and political economy of the city. The new sociologically oriented urban geography vitalised its identity in opposition to what was regarded as the deterministic and outdated 'site and situation' formula, and water, therefore, and everything related to it, was regarded as external to society or as part of the physical environment and therefore outside the range of interest of urban studies.[7]

In the mid-1960s, the 'new urban history' emerged within the discipline of history, being characterised by quantification and model-building aspirations. It gave rise to an influential tradition of urban studies that was primarily interested in the history of urban stratification, spatial patterns (defined sociologically and not by nature, waterscapes or waterzones, etc.) and social mobility. At the same time, what has generally been called the New Left arose as a new political force influencing academia, and encouraging politically and ideologically inspired research on history – from 'the bottom up'. It opened up a very broad topic related to all sorts of social questions. By the end of the 1960s, there was a great deal of interest in urban history, but in line with the political trends it was mainly concerned with what was termed the 'urban crisis' and social relations. The research tradition developed in the wake of the counterculture spreading in most of the Western world and was also inspired by the African-American civil rights movement in US cities in the 1960s and was therefore uninterested in the physical context. Nature or waterscapes were at the time seen as basically non-political issues, and therefore of no relevance to explaining the urban crisis, or even as a hindrance to explaining the things that mattered.

The increasing influence of structuralism led to more studies of cities as structures. The aim became to discover, beneath local differences in surface phenomena, constant laws of an abstract urban culture. This led naturally to a sharper focus on social categories such as class, elites, class struggle, and so on. The various forms of structuralism, whether structural-symbolism which, in its extreme variant, argued that the rules of how the human mind works determine the basic rules and characteristics of urban reality, or structural-historical materialism with its concepts of social formation and modes of production, meant that the role of nature or water in urban development was, if possible, even further relegated to the background. The structures that concerned structuralists were emergent systems of social rules, roles and relationships into which people are born, and which were collectively reproduced and occasionally transformed by human agents, and definitely forms of water-structuration (see below).

As this kind of structuralism went out of fashion, urban studies became more concerned with other issues that could be explained by social variables. The focus now was more on the symbolic and cultural meanings of cities, within the geographical tradition described as a new 'cultural geography' or human geography. The main aim was described as 'unmasking the meaning of cities, landscapes or buildings' (Hall 1998: 28). The city ought to be deconstructed, or 'unpacked' as a text. This opened up a whole new and promising avenue of research. But the studies that were published show that only very few aimed to unpack the 'meaning' of urban water landscapes, or of 'water buildings', such as fountains, reservoirs, river improvement schemes, riverside developments, and so on. Moreover, since one of the most crucial urban infrastructures – the water supply and sewage system – is a truly hidden and literally invisible structure in a strict material sense (because most of it is underground or hidden in walls and in floors), this aspect of the interaction of city and water has naturally been difficult to unpack as text.

The growing influence of post-modernism and radical constructionism in the 1980s and 1990s directed attention further from the impact of nature on social development in general and also on urban development, since they argued that everything originates in the social world, and nature plays a role only to the extent that humans constitute it through their accounts, suspending all forces that are not social or man-made. The influence of nature on urban development was therefore conspicuous by its absence.

When environmental history emerged as a field in the 1980s, one might have expected that this would stimulate more focus on the history of the water-urban nexus, but that did not happen, not the least due to the way water was conceived. The most influential of these new environmental historians, Donald Worster, defined the subject in his lead article from 1990: 'Transformations of the Earth: towards an agroecological perspective in history' (Worster 1990). Environmental history should, according to Worster, be the study of 'the role and place of nature in human life', but the studies of urban environments were explicitly defined as being outside the sphere of interest. The reason was that environmental historians should not study the 'built environment' because this was seen as 'wholly expressive of culture' (Worster 1988: 292–3) and therefore having no significant connection to nature or the workings of nature. Cities were thus defined and reduced to what was seen as 'built environment',[8] that is, as being only a product of social variables, and therefore outwith the interest of environmental historians.

The notion that the 'built environment' is 'wholly expressive of culture' can, however, not be sustained, as shown in the examples of New Orleans and Katarina, Berlin and the Spree, London and the Thames and Beijing and the South-to-North Water Transfer project. The conventional idea that the city and town represent the artificial and built environment that humans make for themselves is a constitutive aspect of urbanisation itself; indeed, to be 'modern' is to live in an urban, built environment. This understanding overlooks the particularities of water, because the built environment one finds in cities is not only a cultural product or a human-made invention. The character of the 'built water environment' will always also reflect the character of the local hydrological cycle or of the physical waterscape, whether the city's water environment is an aquifer, a neighbouring lake or the sea (desalination), or whether the water source is fed by rain, glaciers or is renewable. A nature-centric approach that dismisses the built landscape because it is seen as solely a cultural product, will not acknowledge how nature or waterscapes impact on the built environment. It will not be able analytically to integrate how differences between the built water environment or the water supply systems and sewage systems of New York and Los Angeles, London and Beijing are affected also by different natures or waterscapes. The different control and distribution systems of water that these cities have do reflect different cultures, but they are also the outcome of a complex history of adaptations to different natural waterscapes, efforts at human control and past management ideas of water. The widespread similarities between urban water control systems in many cities all over the world are on the other hand not only the result of similar ideas about water management, but also a reflection of universal characteristics of water as a physical resource.

The three-layered water-system approach provides a way out of this self-imposed cage. It makes it possible to analyse how the urban built water environment is not only a reflection of culture, but also of nature, and both aspects can be integrated in studies of urban development. It can provide a methodology for studying how non-cultural and non-social facts affect how water has been controlled and distributed horizontally, influencing the technology that can be chosen, the material that must be used or can be used, and the size of the water distribution system and how it is operated.[9] The approach can, moreover, link the urban to the non-urban, since urban landscapes impact the hydrosocial cycle far beyond the boundaries of the urban centres, and might indirectly impact on nature or waterscapes in non-urban landscapes far away.

In 2008, the influential *Dictionary of Human Geography* (Johnston et al. 2008) concluded that modern urban geography should deal with all aspects of urban development but one, its physical context, including the local hydrological cycle and waterscape, and hence it had no ideas about how cities impacted on the hydrosocial cycle. The *Dictionary* defines urban geography in such a way that the physical and watery aspect of city development is left outside the scope of interest (Johnston 1980; Carter

1995; Hall 1998),[10] and cities and urbanisms are regarded and studied within a purely 'social world', and seen simply as a response to global and economic trends. In the book *A Theory of Urbanity* (Zijderveld 2009) there is not a word about water, although the author is concerned with the particular culture of cities. The book does not have one single paragraph about the role of rivers, riverbanks, bridges over rivers, mills, cleaning habits, bathing traditions or the role of fountains in urban cultural and artistic history. Textbooks on urban geography share the same approach: the physical context of urban development might be briefly mentioned as a constant and thus 'silent' background factor, but it is basically irrelevant in analysing city developments and the water/urban nexus is of marginal interest since the explicit focus has been on social variables (Hall and Barrett 2012; Knox and Pinch 2009; Knox and McCarthy 2012; Gottdiener and Budd 2005; Pacione 2009; Hartshorn 1980; Kaplan, Wheeler and Holloway 2008).

To summarise: there can be no doubt that the dominant tradition in urban studies has paid scant attention to the universal and structural importance of water in urbanisation processes. Peter Hall, in his acclaimed *Cities in Civilization* (1998), does discuss the role of water in the development of Rome and London, but its index includes no general entries on sewage, water supply systems, rivers, canals or aqueducts. In the same author's book on the future of cities (Hall 2002), the water issue is of marginal interest. A summary of the content of all the volumes of the journal *Urban Studies* between 2006 and 2012 shows that out of 14,363 pages only 86 were devoted to the water issue. These pages were not concerned with the physical or man-made environment as it impacts and is impacted by city development, or with its role in shaping patterns of social activities, power or control. The few articles that dealt with water treated it as a case in studies of political-economic issues, mainly and not surprisingly the water-pricing issue. A total of four articles dealt with such issues. None of them analysed interactions between water systems and cities, and how these impacted on the social and economic life of their inhabitants. The book with the all-inclusive title of *Understanding the City* (Eade and Mele 2002) does not pay the water issue any attention whatsoever. A sociology textbook in an influential series on sociology, *The World of Cities* (Orum and Xiangming Chen 2003: xi), deals only with social aspects of urbanisation, although it claims to be broad and comprehensive in its outlook. The book promises to 'take a journey across time and space, over the urban landscape and to be historical and comparative in perspective'. However, it contains no discussion on the relationship between cities and water whatsoever, nor a single reference to water, rivers, sewage or waterways and canals. Theoretical books on urban politics are either concerned with the urban-water issue and how it shapes both power relations in cities and makes footprints in the waterscape (e.g., Parker 2003; Davies and Imbroscio 2009). The point of departure of this chapter is that modern urban studies have persistently neglected the links between city development and natural and social water systems. The new cultural geography and the textual turn in social sciences, concerned with unmasking the meaning of cities, landscapes or buildings, unpacking it as a text, have generally not been interested in unpacking the meaning of urban-water landscapes. Precisely because water supply and sewage systems are often invisible in a strict material sense, this aspect of the interaction of city and water has naturally been difficult to unpack as a text.

The 1990s saw a rapid growth of urban history in Europe.[11] The European Association of Urban Historians established the *Historical Urban Series* in the 1990s. The series was dominated by a focus on the modern era and a concern with 'urban management' or 'urban governance' (Doyle 2009: 499). None of the more than 35 volumes in the series deals with water and urban history. However, a growing number of books do describe water-urban relationships,[12] and in most popular stories about famous cities the river on which they are located often plays a major role. Many monographs on the history of individual cities deal with the water issue. *Cities and the Making of Europe, 1750–1914*, for example, discusses to some extent the importance both of water supply and sewage and of health issues related to water.[13] But the more overarching questions of how the interconnections between nature and city have influenced the urbanisation process itself as well as the accompanying transformation of the natural landscape, etc. have seldom been raised in a systematic, integrated manner. Although quite a few good narratives of cities and water systems have been published (see, for example, Kelman 2003; Melosi 2008; Bakker 2010; Rinne 2011), we still lack comprehensive, comparative studies that integrate the physical attributes and evolution of cities, and how their history is also reflected in how water has flowed through urban space.

WATER AND URBAN STUDIES – A CONCEPTUAL FRAMEWORK

In spite of the traditions described above, there is a growing understanding of the fact that water-urban interactions are both fundamental and numerous, and that they impact urban economic structures as well as cultural ideas and urban power relationships. But how to analyse these interactions and the water-urban nexuses that are becoming increasingly complex? The research design must be capable of making sense of similarities and differences in urban-water variability contexts, and it must simultaneously be able to integrate water both as a distributor of contamination and as the most efficient urban solvent, as a force of urban destruction and as a source of urban beauty and art, and as a foe and an ally in the fight against urban disease and the complexity of the urban stream syndrome.

One aim must be to be able to narrate the relationships of cities with water coherently, capturing the durability of both their permanence and change. To be able to recognise analytically this wide range of phenomena requires some unifying but also open and non-reductionist frameworks, within which the various elements and their interconnections can be studied. The recognition of the universalism of the particularity of water-urban relationships – cities and water reservoirs, cities and pipe-systems, cities and sewage systems, cities and drinking water, cities and cleaning, cities and health – also requires an analytical framework that makes non-reductionist and comparative, but rigorous, research possible.

The water-system approach suggests a research strategy that addresses the complexity of interconnectedness in ways that can be handled empirically and conceptually. The approach enables analyses that can cater for the fact that water is very dynamic and increasingly multifunctional and historically contingent in both nature and society. It can also account for changes in nature or in the physical waterscape and, especially, for how humans have changed nature and especially the waterscape through history and hence continuously created new possibilities for city development. By employing an analytical framework that covers these different aspects of the urban-water nexus, both nature determinism and social reductionism can be overcome.

The approach suggested here is very different from that suggested by some researchers of urban development who have recently focused on the importance of water but within a mixture of post-modern and early Marxian perspectives. In general, water in these studies is conceived of primarily as a social fact and variable, described as 'streams of power', as a mirror of social development and therefore defined as a hybrid socio-natural phenomenon (Swyngedouw 1999: 445). By underlining that water is 'socio-nature', interesting aspects of power related to water are brought into focus, but the way it is conceived of and emphasised implies that the importance of the earth's water system or the character and limitations and possibilities of urban physical waterscape are of diminished relevance. The local character of the city's physical waterscapes or the confluences between the local hydrological and hydrosocial cycles are of limited or no interest, since it is this distinction between nature and the social itself that the post-modern approach seeks to overcome.[14] The focus of *Water and Society* is that in order to achieve a broad and more complete understanding of the city (including the power struggle over water and how water flows reflects power relations) and its development, it is crucial to describe and understand the enveloping hydrological context as well as the urban hydrosocial cycle and their interconnections. The narrower 'socio-natural' approach cannot explain important variations in hydraulic design, or, for example, London's position as a trading centre on the Thames for hundreds of years before the coming of the steam engine, as a result of the way the river runs in the English landscape. Similarly, the history of Alexandria, founded by Alexander the Great in 324 BC on the very edge of the Nile Delta, cannot be grasped unless the hydraulic design of the past is incorporated into the analysis; that is, how the city could grow because of the way it managed to link two natural water systems, the river and the sea, to the economy at the time. Moreover, the future of both London and Alexandria – as of a host of other cities around the world – is closely linked to potential changes in the physical water system. Authorities fear that if nothing is done, Alexandria will be gradually engulfed by the sea, and that London's future depends on a strengthened Thames Barrier. Similarly, the history of Jericho or Xian or Babylon, or the power struggle behind how New York became the leading city in the USA in competition with Boston, or the particular challenges of modern cities in addressing changing physical waterscapes in the future, cannot be grasped by limiting the perspective to water as 'socio-nature'. One of the important aspects of the history of urbanisation should rather be seen as organising a rearrangement between the importance of the hydrological and hydrosocial cycle in particular urban spaces.

The first layer in this analytical framework for urban studies is concerned with how cities have been and are impacted by the particular, though changing water-urban nexus within which they developed. To study this layer requires knowledge and reconstruction of precipitation and evaporation patterns, river discharges and velocity measurements, aquifers and their behavioural characteristics, that is, knowledge and information about the natural waterscape or hydraulic system that has relevance in the area where the city that is selected for study is located. This layer should be seen as an exogenous, enduring and universal physical factor, but at the same time always having particular characteristics and always being in a state of flux.

The framework makes it clear that there is no direct causal relationship between the physical character of the water system and the location and development of cities, but nor can the importance of the waterscape be reduced to establishing a range of different possibilities. A focus on the physical water system and its importance for city development does not suggest, however, a one-to-one relationship between a certain waterscape and city development, at the same time as all examples of such clear-cut causal connections and correlations should not be rejected out of hand as determinism. The approach should, for example, look for structures, connections and processes between different types of water sources – such as spring water, glacial water, rain-fed water, desalinated water, pumped groundwater – and water control systems, power mechanisms and city development. The approach might also focus on issues like different forms of transport systems that are required to bring water to the city and take sewage away from the city, and how differing water-society systems structure and impact on the opportunitites for urban-water control systems and urbanisation processes in general.

Jericho is a case in point. As early as between 8500 and 7500 BC, it was defended by a stone wall. By the beginning of the third millennium BC, Jericho had become a flourishing city, because it was located around what was later called the 'Ain es-Sultan'. This spring, which in the Bible is known as the Prophet Elisha's spring, provided 4,000–5,000 litres of fresh water a minute without the need for human intervention, and the water was easily distributed by gravity and canals. From the earliest times of settlement until the Ottoman period, this spring was the focus of urban development and decided the location of the city and its main economic activities. The name of the spring indicates clearly that water control was bound up with power relations in the city. Nobody has yet (2014) analysed the whole history of Jericho from a water-system perspective and how it affected the city's relations with the peoples of the surrounding desert, the city's economic activities and social structures and power; no one has as yet followed the flow of water through the city and to the fields for centuries as a method of reconstructing the development of urban power relations.

The highlands of Ethiopia offer an example of yet another locational relationship between city and water. Here rainfall is usually heavy but highly erratic and very seasonal. The location of the country's capital was decided by the presence of a source of water, but not as a dire necessity. Addis Ababa has the healing aspects of bathing as its point of origin. The Queen of Ethiopia, Taytu Betul (1851–1918), had spent much time at thermal springs and requested her husband Menelik I to build a house by a particular spring in the highlands, and the king complied. Soon it developed into a royal settlement, to which Taytu Betul gave the name Addis Ababa, meaning 'New Flower'. The king himself, aware of the curative waters and troubled by rheumatism, established a royal enclosure, a palace and an audience hall at the spring, paving the way for further city expansion. One can still see their remains today on a hilltop not far from the city centre. The location has had far-reaching implications both for the capital's economic role in the country, the building up of its sewage system, and the challenges facing its water system.

Athens undermines the simple, deterministic notion that water's impact on city location has a predictable pattern. Classical Athens was not located where it was because of abundance of water, but rather because of scarcity. According to mythology, there was a competition between Athena and Poseidon regarding who could give the best gift to the city; Athena possessing wisdom and knowledge of arts and crafts, and Poseidon, as the god of water, offering the Athenians a well at the Acropolis. Tradition says that the Athenians voted for wisdom instead of an abundance of water (Koutsoyiannis and Patrikiou 2015), with implications for the city's development up to today but in ways not yet researched. The scarcity of water has obviously impacted the city's economic activities and also the water infrastructure itself, demanding, for example, the utilisation of remote water resources through the construction of the Ilili Aqueduct in 1958 and the Mornos Aqueduct in 1980–1 (Nalbantis et al. 1992: 57–8).

It is also necessary to employ an analytical framework that can incorporate the fact that the physical character of water systems enveloping urban centres is constantly changing, often dramatically. Changes in water systems, due to variations in rainfall, droughts and floods, river velocity and sediment load, etc., impact on urban location and development in many ways. Indian urban history is a case in point. Its history during the millennia cannot be understood without factoring in the relationship between changing water systems and urban development. India is a continent of sluggish, meandering and silt-laden rivers which have given birth to a phenomenon aptly called the 'forgotten cities'. The great rivers of the Indian plain have meandered away, again and again, from cities on their banks (see Chapter 2). When the rivers left the cities, the cities were also left by history. Lacking waterways, they found themselves without the transport routes that had determined the location of the cities in the first place, and they also often lost their significance as ritual centres, due to the importance of holy rivers in Hinduism. The permanence and shifts in this relationship are main reasons for why this physical aspect of the water system cannot be relegated to the background or left in an introductory chapter. A salient aspect of the water-system approach is that it highlights the notion that this structural relationship is neither a one-to-one relationship, nor is it uni-directional, because it is often altered by humans and by nature itself.

The second layer of the water system in relation to urban studies is quite straightforward: one of the features of urban development is that people are stuck with the location of their cities, but that they have gradually become liberated from this locational power of water – by improvements in water-moving technology and in the organisational capacity to control and harness water. All cities change and must change, be retrofitted and re-purposed in relation to their water resources. The enveloping water systems have therefore always been transformed by urban governance and citizenry, and in modern times by the architect and the designer and most importantly by the hydraulic engineer and water manager. It is therefore necessary to employ an analytical approach that captures both how the physical and natural aspects of the local waterscape are underpinning a certain city's location in the first place and produce and reproduce certain possibilities and limitations of city development, and how this water is 'appropriated' and controlled for different demands and reasons at different junctures of its development. The need for water, either for transport, as power generator, as cooler or cleaner, for drinking and production, will always be there in one form or another, at the same time as this need will always change over time, not only due to increases in urban population, but also because changing economic and social activities will put greater stress on the water resources. The waste output of even small cities has also often proven to overtax the absorptive capacity of local aquatic systems. Human modifications are therefore a must for a number of social, economic, cultural and political reasons and all cities in the world have, to different extents and in different ways, modified their natural water.[15] Since the enveloping waterscape of a city is most often part of a wider watershed, a fruitful urban history must also factor in other actors' efforts at using and controlling the waters. The human modifications will therefore also in themselves have varying engineering and hydrological scales and social and political histories. Most cities today are surrounded by an engineered waterscape and a waterscape that is still mirroring, though to different degrees, the character of how the hydrological cycle manifests itself locally in the landscape. This is one reason why the crude dichotomy between nature and culture is not helpful in understanding urban processes, and why there is a need for concepts that on the one hand acknowledge nature's impact but at the same time overcome this way of conceptualising the nature/society divide.

Three examples can illustrate the point: at the time of Nebuchadnezzar II (604–562 BC), Babylon, lying beside the Euphrates on the Mesopotamian flood plain in today's Iraq, was the leading metropolis in the world, measuring about 4.5 square kilometres. Canals from the river had already been built for extensive irrigation, and some of the most impressive were built within the city itself. By the end of Nebuchadnezzar's reign, the eastern parts of the city were also protected by walls which enclosed an area of about 9 square kilometres. Babylon could not have developed where it did had not the Euphrates crossed the flood plain, but human modifications of the water landscape were also needed to develop and sustain it. Mohenjo-Daro in present-day Pakistan was established close to the River Indus but used artificial wells and developed sophisticated systems for supply and sewage. It was one of the major cities of the

Indus civilisation and one of the largest cities in the world in the third millennium BC. In the city, there were possibly more than 700 wells and perhaps as many as 2,000. Each of these had an average catchment radius of only 17 metres, making their density unparalleled in the history of water supply. The most spectacular and well-known water structure in Mohenjo-Daro is the 'Great Bath', a tank measuring 12 by 7 metres, and 2.4 metres deep. However, the use and ritual function of this bath have been more difficult to establish. Another noteworthy development in the city was that it seemed that almost every household had a separate 'bathroom'. But the most intriguing aspect of the city was the system for sewage removal. The physical foundation for this water system was the fact that the water table on the Indus Plain was very close to the surface, so accessing the waters to construct this water system was relatively easy. Modern Rotterdam is a city where water is virtually everywhere, and forms its soul, its identity and its economy. The fight against too much water made this city possible, as the latter part of the name indicates. Rotterdam is a polder city (enclosed by dikes), created by man in a successful struggle to control the water. Rising sea levels and changes in river discharges have forced the city to employ various policies in the course of its history, and the city's development has been totally dependent on the success of the water engineers and will be so in the decades and centuries ahead.

These few examples indicate just how crucial water is for urban history and development and how fruitful it is to reconstruct and analyse the interconnections between these two layers of water systems. The second layer captures everything that humans have done to bring natural water to the city – in all sorts of sectors and for all sorts of purposes. It enables us to describe and understand the water system as an integral part of any city's planning environment. It underlines the fact that this urban-water system, at any point in time the product of cumulative interactions between human purpose and natural waterscapes, both impacts on limits and patterns of action and reflects the economy and technological level of cities. One of the most interconnecting structures in cities is the bonds between people and sectors that these interactions with water produce. An approach that integrates the physical character of water and the structure of the man-made water infrastructure, and the interaction between the hydrological and hydrosocial cycle, enables analyses of linkages that create social cohesion and social hierarchies in particular ways. Analyses that are able to integrate the two layers and the two cycles will encourage investigations of a city's entire 'water machine' as a force and a reflection of urban development. Since the suggested approach includes both the natural and the modified water landscape, analyses can escape being either nature-centric or anthropocentric and this crude analytical dichotomy can be evaded.

The concept of water-urban systems or water-society systems objects to research traditions that perceive urban places and nature as geographical opposites – where cities are regarded as manufactured social creations and nature seen as being irrelevant or non-existent or of limited importance. It distances itself from conceiving nature as something opposed diametrically to the urban. Conventionally the former connotes all that is artificial and socio-technically constructed, and the latter is seen as conjuring up a vision of primal wilderness untouched by human agency. The urban world is in this view defined in terms of what the natural is not. But such a strict opposition is far too simplistic and has long-ranging analytical implications because it is blind to the unique characteristics of water, which undermines such conceived oppositions.

This dichotomic perspective is unfruitful because water is the same both in nature and in the city; water is the same chemical and physical substance when it runs through the most distant forests as when it runs in the tap or in the toilet, and after it evaporates it returns, re-emerging as 'pure, untouched water'. Water's centrality in urban life combined with its unusual capabilities erodes subsequently conventional theories about cities, understood as sites for the conversion of nature by human activity. The water in urban spaces is both natural and socialised at the same time, it is tamed and redirected and in this sense changed but it is still basically water as found in nature. Water in the cities is likewise both a fundamental material structure forming urban processes and a cultural product of the same urban processes, making it clear that definitions of materiality must not be reduced to brute matter or to a static thing since water is always in flux and part of a dynamic social process. The reason why this conceptional dichotomy is rejected here is, as should be clear from what is said above, diametrically the opposite of why social theorists like Latour and Beck object to distinctions between the natural and the social. They claim that nature should be reduced to a social construct, while the water-system approach which is concerned with the interactions and confluences between the hydrological and hydrosocial cycles, maintains that it is crucial for our understanding of urban processes that nature should not simply be reduced to a social construct. The water-system approach objects to the current conventional argument that pure nature does not exist any more. Since water in the hydrological cycle is reborn as pure nature continuously after being socialised, and large parts of the planet's water bodies are not affected by the hydrosocial cycle at all, one cannot suggest that water has become social. The approach suggested in this book also distances itself from the viewpoint that the distinction between the natural and the social should be rejected because it is often difficult to delimit the imaginative and material boundaries between what is understood as natural and what is understood as urban. The water-urban system approach argues rather that it should be maintained but reconsidered in order to cater for the fact that water itself objects to the distinction in a very particular way since it is the same or has the same characteristics and capabilities in urban centres as in the wildest nature.

The third layer of what is here in general called the water-system approach captures the 'cultural' dimension of urban water. By integrating this dimension in the analytical framework it will make it possible to analyse the linkages between cities' specific physical waterscapes and how it will always be filtered through varying cultural lenses, the modified and controlled water resources that exist at a certain time, which will always reflect actors' ideas of water and how it can be used in the past, and those conceptions and managerial plans regarding water and city development that are developed to handle the urban-water nexus, some of which will be implemented while others will lose out. This analytical layer recognises and encourages analyses of how water both in nature and in society, and as a natural resource and as a social good, will always be culturally constructed and filtered, differently by different actors, and from time to time and from place to place. A comprehensive history of the ideas of water has not yet been written.[16] A comprehensive study of urban ideas about water would have thrown new light on city development.

It is necessary to focus on ideas about water in relation to a broad range of issues: as a means of exerting urban social and cultural power, as an object of management practices, as a religious and cultural symbol or object, as an image of the natural in the urban world made up of concrete, as a place of worship or of social gathering, and as a signifier of social and cultural distinctions. By being linked directly to the two other analytical layers it will highlight the importance and permanence of notions and practices about urban water management. By studying these ideas and their development, based on the premise of the increasing multifunctionality of water, it becomes possible to capture how these ideas have changed radically over the years, how they differ from urban centre to urban centre, and how different social groups in the same cities often have very different ideas about water and water control.

Ideas about water will obviously differ in Cologne and Manchester – two of the cities with the most rainy days in Europe – from those in Wadi Halfa in Northern Sudan, where years might pass without a raindrop and where the Nile brings all the water the people have under a permanently blue sky. These 'cultural' differences in the ideas of water reflect different notions of space and place, but cannot be reduced to a social construct only, since they are underpinned by differences in waterscapes and water-society relations. Ideas about management of water will also be very different in Mexico City, where around 20 million people must be provided with water, much of which is lifted more than 1,000 metres, utilising seven reservoirs, a 127-kilometre-long aqueduct with 21 kilometres of tunnels, 7.5 kilometres of open canal and a water treatment plant, from those in Beijing, which gets much of its water from a canal 1,800 kilometres long bringing the life elixir from the Yangtze River in the south, or Bergen in Norway, where there is an annual rainfall of more than 2,000 millimetres. The role and character of these types of ideas are a neglected field of study, since such research has to combine non-social and social variables and how they interconnect over time and according to space and place.

An analytical approach that firmly integrates research on ideas of water and their relevance to urban processes will also be able to throw new light on urban power struggles, since unfolding economic processes involving particular forms of water valuation and commodification and differences in how water is understood are behind one of the most conflicting issues in the contemporary urban world: should water be seen as a normal market commodity, as a common good, or as a human right? Even in the most modern cities ancient myths and religious ideas about water can be found closely interlinked with modern-day nostrums about water as a natural element and in relation to its many functions for the biosphere and for humans in particular. This battle between different ideas about water will impact on urban development in the future, since it will influence whether water can be priced and thus the provision of urban water infrastructure.

This layer focusing on ideas of water highlights the different ways in which water is socially constructed and will integrate into the analysis of how water has been conceived differently in cities over time and in different cities at the same time, showing both the endurance and the instability of meaning and the coherence and fragmentation of habits of thoughts when it comes to water and water control. Linking the ideas of water to the two other layers of the water system allows this production of cultural metaphors to be analysed in both a material and a social context.

One contemporary example that can illustrate the importance of ideas of water and how they change and frameworks for urban developments is the growing emphasis many urban dwellers put on the importance of free-running water. In certain urban places, free-running rivers have acquired precious cultural and spiritual value. Daylighting rivers, that is, uncovering buried rivers that used to be channelled directly into culverts and underground sewer systems, has become a political movement, and will impact on future urbanisation processes. Such ideas have far-reaching consequences for urban development since their managerial implications will have an immediate impact on the basic urban infrastructure and also on social relations.

Ideas about water in cities have influenced social and economic life in the past, and still do in the present-day world. Water acts as a symbol of man's power over nature in the form of fountains (the fountain in the desert city of Phoenix where water flies many metres into the sky is an interesting case), water as a means of urban purity and health (all the baths and fountains of Rome), and water as a central expression of urban power and distinctions (the Pope's fountain on Piazza Navona, Rome). How the water runs through a city at any point in time can therefore also be treated as a mirror of urban development in general.

The concept 'open, complex and multifunctional water-systems' treats water and the city as being intertwined, bringing into focus how cities use and control their particular water at every level of urban activity, and in turn affects water in a myriad of ways as it runs through the cities. It underlines and gives a framework for analysing how the flow of natural and social water through urban space is a permanent trait of any city's development at any point in history and all over the world. From the intimacy of the modern bathroom to the buried space of the sewer or the underground aqueduct, from the thrill of the drops in a fountain to the tamed but powerful water behind dams and reservoirs, water has provided a link between the concrete and worldly experience of space and both the material and immaterial dynamics of urbanisation. Physical waterscapes and modified waterscapes also reflect and exhibit a diversity of responses to urbanisation. More or less identical urbanisation processes may have very different impacts on the enveloping waterscape also because these same waterscapes differ in their abilities to satisfy urban development. The concept encourages research that can describe and analyse how the same water plays multiple roles in the lives of cities: it has caused disease, squalor and human misery, and provided the means to battle these same urban problems.

The water-system approach can capture how human-modified water systems in their turn change the physical water system in an everlasting process of mutual interaction. It takes as a starting point the fact that no urban water landscape is either completely natural or completely controlled or socialised, because urban development presupposes on the one hand modification of the natural waterscape, but is on the other hand not able to destroy water as nature although the actual water source may be depleted or polluted. Efforts to control water and the built water environment it creates must be analysed not only as a reflection of 'culture', but also as being impacted by the physical character of the waterscape. The actual flow of water in cities must therefore be analysed as being located within both a particular physical context and a particular set of traditions related to water, which in its turn is located of course within broader political relations and social rhythms.

This chapter has suggested an approach to urban studies and the history of cities that is multidimensional and that integrates nature–society relations. The original aspect of this approach is that it enables analyses of the physical, social and cultural aspects of the water-urban nexus and their inter-relationships. The approach is also useful because on the one hand it emphasises that the empirical links between a city and its water system are always particular, while on the other hand the three layers of the water system are universally applicable as an analytical framework because all cities connect to these layers of the water system, but importantly and that is why it becomes so fascinating, in very different ways, in order to become a city. There is an urgent need to broaden and open up urban studies, making them less reductionist by developing analytical approaches that both theoretically and methodologically integrate physical and social factors of relevance to city development. This will be crucial to obtaining a fuller understanding of how cities have developed through history, but due to the extremely rapid and global urbanisation process and the growing uncertainty about the water of the future it is also an issue of great current political and economic urgency.

6

WATER, SOVEREIGNTY AND THE MYTH OF WESTPHALIA

This chapter will focus on the issue of sovereignty from a rather unusual perspective. Sovereignty has for centuries been at the very centre of political and legal arrangements. It has been one of the constituent ideas of the post-medieval world, and it is the central organising principle of the system of states in the present-day world. The meaning and changing nature and status of state sovereignty in international politics and law have been analysed in innumerable articles and textbooks. Despite this, it is still widely regarded as a poorly understood concept, a confusion stemming from different sources. The sovereignty doctrine has 'been turned inside out and upside down by the successive uses to which it has been put', it was argued already in 1928 (Ward 1928: 168). The doctrine has in line with this been cited as authority for acts never intended as expressions of sovereignty, and it has been refuted in forms that never existed in the real world. It is a term that in the contemporary world extends across continents, religions, civilisations, languages and ethnic groups, and different constructs of the sovereignty concept exist, offering varying and contradicting answers to the question of what it is.[1] But most scholars agree that at its core, sovereignty is typically taken to mean the possession of absolute authority within a bounded territorial space: 'A sovereign state can be defined as an authority that is supreme in relation to all other authorities in the same territorial jurisdiction, and that it is independent of all foreign authorities' (Jackson 2007: 10). It is this notion of the centrality of territoriality, which makes the question of sovereignty so interesting from a water-society perspective.

Here sovereignty will be analysed in relation to how state actors have performed when it comes to international rivers and aquifers, and how interactions with this particular fluid web of nature have impacted notions of and practices of sovereignty. Within this general framework we believe it is most fruitful to focus on three specific areas or central topics in the international discourse on sovereignty: (a) what was the Westphalian notion of sovereignty; (b) to what extent has history been a development from a Westphalian to a post-Westphalian notion of sovereignty; and (c) what are the connections between sovereignty and conflict. We will show that in international relations studies and in international law these issues will appear in a new light by focusing on these three confluences between international rivers and politics.

THE MYTH OF WESTPHALIA AND CO-OPERATION OVER WATER

A main assumption and premise in the very extensive legal and political science literature is that the idea and principle of sovereignty is a legacy of the Peace of Westphalia in 1648. Westphalia is seen as the very birthplace of the idea of absolute and unrestricted sovereignty. The main story goes like this: the Westphalian model emerged against the background of the cataclysmic changes unleashed in Europe during the sixteenth and seventeenth centuries, partly as a result of the Reformation. The peace agreement of 1648 provided legitimacy for the principle and idea of the territorial, unitary and absolute sovereign state. Through the centuries after 1648, this legacy and ascribed tradition – as theoreticised by political scientists – increasingly emphasised sovereignty, and led to confrontation of claims of absolute territorial sovereignty with claims of the absolute integrity of state territory. Westphalia has come to symbolise the birth of a new world in which states are nominally free and equal and enjoy supreme authority over all subjects and objects within a given territory, engage in limited measures of co-operation and regard cross-border processes as a 'private matter' (see Falk 1969; Held 1995: 78; and, for quote, see Held 2002: 4).

In the last decades there has been a big debate about whether we live in a post-Westphalian world or not. One 'school' argues that due to a number of global trends, the triumphant Westphalian notion of sovereignty is now gradually undermined. It is claimed that we live in a post-Westphalian age (Harding and Lim 1999; Westra 2009; Macqueen 2011) characterised by the 'end of the sovereign state' (Wunderlich and Warrie 2010: 256). Other researchers question the realism and validity of this claim, arguing that international relations remain anchored to the politics of the sovereign state (Buzan, Jones and Little 1993). They hold that differences in national power and interests, not international norms of co-operation and supranationality, continue to be the most powerful explanation for the behaviour of states. Both these 'schools' agree, however, that Westphalia signalled that the idea of the sovereign state having a final and absolute authority over its territory was born and subsequently became dominant.

We will argue here that if Westphalia really marked the triumph of unfettered sovereignty, then a study of the history of the intricate and long negotiation process and analyses of existing transboundary linkages should support this description.[2] First, we will take a close look at the original texts.

Article I in the agreement reads like this according to an internet edition published by Yale University:

> That there shall be a Christian and Universal Peace, and a perpetual, true, and sincere Amity, between his Sacred Imperial Majesty, and his most Christian Majesty; as also, between all and each of the Allies, and Adherents of his said Imperial Majesty, the House of Austria, and its Heirs, and Successors; but chiefly between the Electors, Princes, and States of the Empire on the one side; and all and each of the Allies of his said Christian Majesty, and all their Heirs and Successors, chiefly between the most Serene Queen and Kingdom of Swedeland, the Electors respectively, the Princes and States of the Empire, on the other part. That this Peace and Amity be observ'd and cultivated with such a Sincerity and Zeal, that each Party *shall endeavour to procure the Benefit, Honour and Advantage of the other* [my italics]; that thus on all sides they may see this Peace and Friendship in the Roman Empire, and the Kingdom of France flourish, by entertaining a good and faithful Neighborhood.[3]

In an English translation from 1697 it reads slightly differently:

> That there shall be a Christian and Universal Peace, and a Perpetual, True, and Sincere Amity, between the Sacred Imperial Majesty, and the Sacred Most Christian Majesty; as also, between all and each of the Allies, and Adherents of the said Imperial Majesty, the House of Austria, and its Heirs, Successors; but chiefly between the Electors, Princes, and States of the Empire on the one side; and all and each of the Allies of the said Christian Majesty, and all their Heirs and Successors, chiefly between the most Serene Queen and Kingdom of Sweedland, the Electors respectively, the Princes and States of the Empire, on the other part. That this Peace and Amity be Observed and Cultivated with such a Sincerity and such Zeal, that each Party *shall endeavour to procure the Benefit, Honour and Advantage of each other* [my italics]; that thus on all sides they may see this Peace and Friendship in the Roman Empire, and the Kingdom of France flourish, by entertaining a good and faithful Neighborhood.[4]

In the original French text it reads like this:

> & cette paix s'observe & cultive sincerement & sérieusement, enforte que
> chaque Partie procure l'utilité, l'honneur & l'avantage l'une de l'autre,
> & qu'ainsi de tous côtés on voye renaitre & resleurir les boens de cette
> paix & de cette amitié, par 'l'entretien sur & reciproce d'un bon % fidele
> voisinage avec [...]. (Bougeant, vol. 6: 285)

What clearly emerges is that these texts were not – contrary to received
wisdom – a treatise for absolute sovereignty, but underlined the value
of restricted sovereignty and a concern for the interest of each other.
The above English and French versions of the text of the peace treaty
underlining the principle of the 'interest of each other' or 'of the other'
falsify assumptions that the Peace of Westphalia did establish the
principle of unrestricted sovereignty. In spite of the countless books and
articles stating the opposite, the text of the peace agreement formulated
and reflected ideas of common benefits.

What is of specific concern in this connection is how the peace
agreement described the role of the transboundary rivers in relation to
territorial sovereignty. Westphalian sovereignty has often been described
as a concept of the sovereignty of nation states on their territory, with no
role for external agents in domestic relations or structures, and Westphalia
is seen as the place that ended attempts to impose supranational authority
on European states. But what did the agreement say, and here we limit
our attention to the River Rhine due to its importance.

Paragraph LXXXIX of the agreement deals explicitly with the River
Rhine:

> All Ortnavien, with the Imperial Cities of Ossenburg, Gengenbach,
> Cellaham and Harmospach, forasmuch as the said Lordships depend on
> that of Ortnavien, informuch that no King of France never can or ought to
> pretend to or usurp any Right or Power on the said Countries situated on
> this and the other side of the Rhine: nevertheless, in such a manner, that
> by this present Restitution, the Princes of Austria shall acquire no more
> Right; that for the future, the Commerce and Transportation shall be free
> to the Inhabitants on both sides of the Rhine, and the adjacent Provinces:
> Above all, the Navigation of the Rhine be free, and none of the Parties
> shall be permitted to hinder Boats going up, or coming down, detain,
> stop, or molest them under what pretence soever it may be, except the sole
> Inspection and Search which is usually done to the Merchandizes, and it
> shall not be permitted to impose upon the Rhine, new and unwonted
> Tolls, Customs, Taxes, Imposts, and other like Exactions.[5]

The text of the agreement underlines the importance of co-operation and the need to restrict the absolute territorial power of the sovereign; that is, the opposite of what has been generally said about it. But was this only a 'slip of the pen' and could it therefore simply be discarded as irrelevant in any analysis of how Westphalia understood sovereignty?

If the peace agreement's plan for the Rhine is analysed in a broader historical and geographical context it becomes clear that the text was not an accident, but that it reflected new and emerging ideas about how the countries on the Continent could benefit by improving rivers and waters to promote wealth and trade.[6] The importance ascribed to supranational co-operation over waters in the peace agreement was the combined effect of the hydrological and geographical character of the continental rivers in an era when the use of the river was primarily for the transport of goods, and a deliberate economic strategy pushed by leading architects of the peace process.

The Rhine, with a basin of about 180,000 square kilometres and a length of 1,300 kilometres and comprising what is today the northern tip of Italy, Switzerland, Austria, Germany, France, Luxembourg, Belgium and the Netherlands, was (and still is) one of the most central trading routes in Europe. The Rhine posed in its natural state many hazards for navigation and thus for trade, even for quite small vessels.[7] From Roman times, attempts had been made to improve particularly awkward stretches of the river, but success had been limited. In the centuries and decades before the Peace of Westphalia nothing much had been done in order to improve it. The river's nature created new obstacles incessantly. The river frequently shifted its course in floods, sometimes leaving flourishing river quays stranded. Towpaths and dikes were destroyed. Rocks and reefs impeded shipping.

In Germany in the early Middle Ages commercial shippers ran scheduled trips along the Rhine between Mainz and Cologne. Although the medieval records fail to establish precise quantitative data about the volume or value of riverine traffic and trade, it is safe to assert that trade was vital though still limited. On an average, all-year basis half of the water came down from the Alps (mostly in spring) and half from the tributaries north of Basel (mostly rain-fed). The water sources of the river thus liberated the Rhine from some of the problems encountered in other French and German rivers. The fluvial dynamics of the Rhine above Strasbourg prevented, however, the construction of permanent towpaths and forced upstream traffic to depend on human muscle power or wind. Upstream travel was very difficult, requiring towpaths and the change of ships frequently on the way. The Rhine's 'low-to-high flow ratio' coupled with the föhn winds, meant that the river was flood-prone. The

Upper Rhine had the classic characteristics of a floodplain, and frequent floods made quay building and the development of a trade infrastructure hazardous enterprises. Catastrophic flooding happened in 1124, 1342 and 1573.[8]

Traffic on the Rhine suffered for natural and hydrological reasons, but also because of political boundaries. Prior to the Thirty Years' War, the river was under the control of the Emperor of the Holy Roman Empire and the imperial princes were responsible for maintaining the navigability of the river. The princes' authority was weak and they were more concerned with extracting tariffs for themselves than using resources to improve the river. The town guilds along the river acted in the same manner. In the mid-seventeenth century, kings, bishops, cities and robber knights tried to profit from Rhine navigation. There were numerous tolls along the Rhine and passing ships had to pay duties to the rulers of the different Rhine sections. The number of tolling stations had increased from 19 in the late twelfth century to over 60 by the sixteenth century (Mellor 1983: 70). The way the Rhine ran through the landscape made it quite easy to control the trade on the river, as signified by all the castles that were built along the riverbanks and that can easily be seen from the decks of the cruise ships floating down the engineered Rhine of today. The taking of tolls was held to be an imperial right and liberal grants were made to cities and especially lords to secure loyal support for the Emperor, or as a means of filling an empty treasury. There was, moreover, no point in an individual prince improving his stretch of the river, if the other princes did not do the same along their stretches because individual action would not improve it either for the individual prince or as a common good.

The Peace of Westphalia was among other things an effort to do something with this potentially very useful north–south routeway for transport through continental Europe.[9] In spite of all the problems with river transport on the Rhine, it was still considered the preferable way to move goods and passengers. Previous rulers had occasionally tried to eliminate the tolls by force but these attempts had failed. One fundamental aspect of the diplomatic and economic strategy of the French Cardinal Jules Mazarin (1602–61), the man who effectively ran the French Government during the Congress of Westphalia, was his vision for the continental waterways. His aim was to weaken the authority and power of the Emperor and one way that France could achieve this was to facilitate economic development in the German states. The best bet was to improve the waterways, since improved trade on the rivers would also benefit France. He thus commissioned a study of the rivers of the European Continent and the potential for an expansion of the trade of goods produced along these rivers: the Vistula, the Oder, the Elbe, the Weser, the Ems (which crosses Westphalia) and, of course, the Rhine, the dominant economic channel linking Switzerland, Germany, France and the Netherlands.

The political and territorial system on the Continent hindered the development of the Rhine as a trading artery and thus Mazarin's strategy for weakening the Emperor. He saw the Rhine as a corridor of development, but it was misused by the princes, working against their best interests. In 1642, France announced that there would be no further peace negotiations if it was not forbidden to create new tolls along the Rhine River. Even though the edict was not implemented in full, it was crucial for creating the political atmosphere that enabled the Congress of Westphalia to succeed. The edict was seen as an important economic and political initiative, benefiting not only France but the whole region since the river was a key trading route on the Continent. The understanding was that the economy was devastated by war, but was further undermined by the burden of systematic interruptions of trade on the river between Northern and Southern Europe.[10] Legally, the use of the rivers was regarded as a common right and the use of the water for drinking and voyaging was free, thus undermining the idea of absolute territorial sovereignty. Grotius argued that duty could not be taken for the exercise of this right, but that it should be interpreted as a compensation for the cost of maintaining the duty and the towpaths. The Frankish monarchy on the other hand saw it as a tax upon the river, rather than a denial of the right of passage (Chamberlain 1923: 146–7).[11]

The agreement did not lead to fundamental improvements of the river as a trade route. Westphalia did not solve the problem of the Rhine. The regime on the Rhine in the eighteenth century has rightly been characterised as a 'landscape of petty quarrels'. Between Alsace in the south and the Netherlands in the north there were 97 German states alone. The 'knights and priestlings' ruling these tiny states were warring with their neighbours over fishing holes and bird islands. They built some small dams, with only local aims in mind, and these only increased the number of sandbars and forks. They of course defended their 'staple' and 'transfer' privileges, an important source of income to them, and manned the toll booths (34 in a 600-kilometre stretch from Gemersheim to Rotterdam alone) – all negatively impacting on river trade.

The Treaty of Westphalia regarding the Rhine coupled with the ideas about restricted sovereignty and co-operation over the Rhine can be seen as the first formal germs of what later – in 1815 – became the pioneering Rhine treaty in the history of European co-operation and unification and in international water law. The situation was somewhat improved but the problem was not solved and elimination of tolls was therefore also an important issue in the peace conference in Vienna in 1815, after a number of agreements had been signed in the years before, such as the

Treaty of the Hague in 1795 and the Convention of Paris in 1804 on the tolls on the navigation of the Rhine. In the framework of this peace treaty, the riparian Rhine states voluntarily voted for free navigation and elimination of the tolls. They created the Central Commission for the Rhine navigation. The internationalisation of shared rivers and lakes for navigation was initiated formally in 1815 at the Congress of Vienna when the Rhine Commission was established, then the Oder and the Niemen in 1918, the Elbe in 1921 and the Weser in 1923 were all declared international waterways for navigational purposes. In 1856, the Treaty of Paris internationalised the Rhine and the Danube. Later, the development of hydropower led to the 1932 Geneva Convention for the development of hydropower in rivers affecting more than one state. The Peace of Westphalia can in this water perspective not be seen as belonging to a political tradition of unrestricted sovereignty and a tradition which is currently undermined due to present economic developments and ecological constraints. On the contrary, by viewing regional development as a historical process there is a clear connection between 1648 and 1815 and 1932. The principle of sovereignty was from the very beginning modified by the rationality of and the need for co-operation over international waters.

This short assessment of what took place regarding the waters of Europe is sufficient to falsify the dominant interpretation of Westphalia.[12] To assert that it was Westphalia that was 'formally recognizing exclusive territorial jurisdiction of monarchs' (Wunderlich and Warrie 2010: 255), that it was in 1648 that the idea of undivided, unlimited authority and territorial exclusivity was born, contradicts not only the development of the actual peace process and the role of transboundary waters but also the text of the peace agreement. The text underlined rather the need for considering the interests of 'the other'. It also prescribed co-operation over the Rhine that ran through different sovereigns' territories in the following words: 'the Navigation of the Rhine be free, and none of the Parties shall be permitted to hinder Boats going up, or coming down, detain, stop, or molest them under what pretence soever it may be, except the sole Inspection and Search which is usually done to the Merchandizes, and it shall not be permitted to impose upon the Rhine, new and unwonted Tolls, Customs, Taxes, Imposts, and other like Exactions.' The concept of sovereignty as understood in 1648 meant that an aspect of being a member of an international society of states was the requirement to comply with international agreements and to contribute to and participate in the solution of collective problems. The absolutist definition of sovereignty cannot therefore be historically justified in the way it has been, and the canonical story is in this sense a myth. But like all other myths in history, it is a myth for a reason: it has served specific political and ideological interests. The mythical story should therefore be analysed as yet another expression of the political-ideological career of the notion of sovereignty. Already in 1928, it was described in this way; 'the various forms of the notion have been apologies for causes rather than expression of disinterested love for knowledge' (Ward 1928: 167).

THE MYTH OF 'POST-WESTPHALIA' AND THE CONFLICT OVER WATER

The dominant assumptions about the gradual undermining of the idea and doctrine of sovereignty form the backdrop of statements about the 'death of Westphalia' – a widely used metaphor to capture the perceived fall of status and strength of the sovereign state.

This assessment of the gradual decline of sovereignty has been forwarded by an influential school within international relations studies. In the 1970s and early 1980s, liberal interdependence theorists (Keohane and Nye 1972 and 1977; Morse 1976; Rosecrance 1986) argued that due to global development trends, state sovereignty was being eroded by economic interdependence, global-scale technologies and democratic politics. The sovereignty of states was within this perspective more and more constrained and penetrated by 'the forces of globalisation', of which international organisations can be thought to be a part (Litfin 1997). There was a shift away from state-centric to multilayered global governance (Held 1995).

Much literature has argued that transnational environmental interdependencies have led to the demise of the state system. The ascribed mismatch between what has been conceived as the requirements of physical ecology and reality of the social structure of politics has been expressed most famously, perhaps, in the dictum of Our Common Future: 'The Earth is one, but the world is not' (World Commission of Environment and Development 1987: 1). Some have anticipated its eventual replacement by some far-reaching supranationalism or even by world government (Falk 1969; Ophuls 1977). A number of scholars and activists have argued along the same lines that the earth itself demonstrates the inadequacies inherent in legal principles based on states' territorial sovereignty. It has been assumed that the cumulative impact of agreements on ecological issues would tend to undermine the institution and idea of state sovereignty, since the territorial exclusivity upon which state sovereignty is supposed to be premised appears to be fundamentally violated by transboundary environmental issues (Johnston 1992). Or in other words, the seamless web of nature is seen to be contradicting and eroding the man-made system of territorial states and therefore also the doctrine of state sovereignty.

Based on the above premises the following hypothesis could be formulated: since water is always in flux, constantly neglecting political and cartographic territorial boundaries, it should be assumed that this trend is particularly visible in the management of international river basins due to the increasing significance for water resources management of regimes of supranational governance. International rivers should by their very nature be constantly undermining the idea of sovereignty.

At first glance, the hypothesis is confirmed. Legal theories of 'absolute territorial sovereignty', according to which a state has an absolute right to do as it pleases with the water in its territory, and the theory of 'absolute territorial integrity', whereby the riparians are considered as having an absolute right to the natural flow, unimpaired in quantity and quality, are not recognised in the contemporary world. But this development cannot merely be interpreted as a sign of such a trend. Sovereign rights to utilise the water have for a long time and in many societies been limited by the obligation to consider the sovereign rights of other stakeholders. As we have shown, that was already an aspect of 1648 as it was of the Rhine and the Danube conventions. The first agreements about the Nile from the 1890s and the first decades of the twentieth century barred upstream countries from using the Nile without the consent of other states in the basin.

There has been a noticeable growth in the number of international institutions – both UN-sponsored institutions and NGOs – and basin-wide organisations that have made 'sovereignty bargains' an art of politics executed by many state actors. In some geographical areas one can discern a development in which states sharing international water resources have moved from positions based on notions of unrestricted sovereignty to positions which recognise the need to limit their sovereign discretion on the basis of sovereign equality. But this development does not necessarily mean a weakening of the sovereign. As it has been argued, states may engage in sovereignty bargains in which they 'voluntarily accept some limitations in exchange for certain benefits' (Litfin 1997: 170). If that is so, this development does not entail a weakening of sovereignty, just a change in the form of its manifestations.

There are examples that indicate that the assumed trend towards a weakening of the idea of sovereignty in relation to international river basins is not so clearly directional. Parallel to the internationalisation of water politics and water management, the status of the notion of sovereignty has been strengthened 'on the ground' in many parts of the world. The post-colonial history of the management of the Nile is a case in point. An increasing status of state sovereignty has to a considerable extent been developed in relation to questions about how this international river in the age of modern technology should be managed among different 'stakeholders'. In recent years, countries like Kenya, Uganda, Rwanda, Tanzania, Burundi and Ethiopia have demanded as sovereign states the right to use the waters of the Nile running through their lands down to Egypt and the Sudan, rejecting agreements entered into by the colonial rulers. The negotiations over the use of this common resource have created a very important and new arena for these states to demonstrate their sovereignty. The Nyerere doctrine of the 1960s and Kenyatta's proposal put forward at the same time were crucial initiatives and steps in the history of exercising state sovereignty in the region (Tvedt 2012). Similarly, the Nile Waters Agreement between Great Britain and Egypt in 1929 was a watershed event in Egypt's march towards a sovereign state after it gained formal independence in 1922, just as the Nile Waters Agreement between the Sudan and Egypt in 1959 signified the Sudan's emergence as a sovereign actor in the world scene and in the regional scene. The prolonged discussions between India and Bangladesh about the Ganges and the Farakka Dam have, if anything, made the status of state sovereignty stronger and increased the animosity between the two neighbouring countries. The problems inherent in sharing international aquifers show some of the same development. The discussions among the countries with territories covering the Guarani Aquifer in South America and in the International Law Commission's 2008 Draft Articles on Transboundary Aquifer that in 2010 led to the Guarani Aquifer Agreement have strengthened the status and relevance of state sovereignty. The agreement asserts that water resources *belong* to the states in which they are located and are subject to the *exclusive* sovereignty of those states. These cases falsify the overall hypothesis about a universal, historical trend, and ask for more detailed empirical research.

The actual historical development is more multifaceted than the dominant trend analysis, but why is this so? Since sovereignty is not only an attribute of the state but is attributed to the state by other states or state rulers, there are no reasons why international or transnational river basins or aquifers should – due to a kind of geographical necessity – erode the status and legitimacy of sovereignty. It turns out that geographically unnatural borders across international water bodies are challenged by international institutions and modern legal thinking, but that they also serve an increasingly important symbolic function in encouraging manifestations of state sovereignty. By focusing on territorial borders within a river basin the political leaders make themselves easily visible as defenders of 'the interests of their people'; since all inhabitants in all states need water, the state and its leaders can acquire legitimacy as sovereigns defending their people in negotiations over such cross-boundary ecological structures.[13] States exercise this sovereignty, moreover, often in multilateral, international institutions, characterised by being distanced from societal and democratic control since state bargaining with society is bypassed and also normatively defended by the idea of multilateralism. This externally induced, state-led challenge of democratic control should not simply be interpreted as an erosion of sovereignty. It might rather be a sign of an opposite development, since this context gives the actors representing the sovereign increased freedom in their sovereignty bargains. This is especially so in relation to international waters, where it is easy to use nationalistic slogans to mobilise people in the street but where more de-politicised negotiations may be the most optimal solutions both for the river and for the states sharing it. To use a reified, ahistorical notion of sovereignty, disregarding the actual complexity of practices that exist, will fail to grasp the multiple dimensions of sovereignty and its meanings and how these are in constant flux.

This short analysis has rejected the universal validity of the above trend analysis of the status of sovereignty, primarily by testing the hypothesis in relation to international river basins, an area where one should assume that its validity should be confirmed.

The dominant but mythical story about Westphalia misrepresents the past with the consequence that the present is misunderstood: the differences between 'then' and 'now' are far fewer than the talk about 'the end of Westphalia' presupposes. Regarding water and river management in particular it is empirically misleading and theoretically problematic to talk about a post-Westphalian age signified by co-operation and the undermining of the absolute power of the sovereign, since Westphalia initiated an era of co-operation over water between sovereigns in continental Europe.

SOVEREIGN STATE ACTORS AND CONFLICT AND CO-OPERATION OVER WATER

The manner in which states conduct their hydropolitics with one another has in general within the field of international relations been analysed through theoretical frameworks associated with ideas of the sovereign state actor (Dalby 1998). The basic idea within this tradition is that unilateral development based on sovereign's interests will be conflicting by nature. We will here contest this general assumption. We will argue that the idea shared by realists and neo-realists – that sovereign states driven by power and interests will find it very difficult to co-operate given that they ultimately insist on maintaining and safeguarding their own autonomy, control and legitimacy – overlooks both the nature of rivers as transboundary resources and how these can be approached by state actors. Empirical, historical studies show that neither conflicts nor 'tragedies of

122

the commons' are given or guaranteed outcomes in the absence of co-operative framework agreements in international river basins. This is so due to a combination of natural characteristics of water bodies and historical processes, issues seldom integrated within this type of social analysis (Tvedt 2004a and 2010).

Given the fact that almost all big rivers are shared by two or more sovereign states and that almost half the population of the world lives in international river basins on the one hand, and that there have been very few wars or open conflicts between sovereign states about how to use these rivers on the other hand, the sovereign states have managed to solve a lot of conflicting issues in a peaceful manner. This is not an argument against the idea that sovereign state actors create conflict, but it refutes the assumption that they are not able to solve differences in a peaceful, non-conflictual manner. It is possible to regard the unique co-operation among sovereign European states over the Continent's big rivers in the seventeenth and early nineteenth centuries as a forerunner to the European Union of the twenty-first century. The Indus Water Treaty in 1960 was made possible by an agreement between two sovereign states, brokered by the World Bank and disregarding the interests of individual regions, such as Kashmir, and ethnic groups in the basin. There are thus a number of examples that show that states can enter into agreements and by such an act contain potential conflicts between other and different actors.

But there is another geographically related argument that is more interesting and intriguing (see Tvedt 2010 and 2014) when it comes to the role of the sovereign. In large river basins, economic, political, technological and ecological conditions can vary extremely from one part of the basin to another, and this fact presents sovereign states located within international river basins with different and potentially non-conflicting strategic choices. Climatic conditions, soil types, velocity and route characteristics may have created fundamentally different options of adaptations in the past, and they will create a wide array of possibilities in the present and the future. For example: irrigation may not be a priority in one country as a sufficient amount of rainfall enables rain-fed agricultural production there, while, at the same time, irrigation may be a fundamental approach to water resource utilisation in another country within the basin. In some parts of the basin the water can produce hydropower, while in other parts this is not possible. The need for water and the way states relate and are capable of relating to it will vary markedly depending on a number of historical factors. The point is: the sovereign's territorial interests in maximising water usage may not always be in conflict with another sovereign state's plan to do so, contrary to what is the case in a traditional 'commons' as described by Garrett Hardin and others. Even in cases when sovereigns enjoy full sovereign freedom, their actions may be to the benefit of others. When, for instance, Ethiopia erects the highly controversial Grand Renaissance Dam within its territory, it might frighten Egypt for geopolitical reasons, but it may still be in the real, long-term interest of Egypt as far as technical management of scarce water resources is concerned. In an international river basin it must not always be a zero-sum game, where a participant's gain (or loss) must be balanced by the losses (or gains) of the utility of the other participant(s). Instead of having a situation where when the total gains of the participants are added up and the total losses are subtracted, the result will be zero, one will have a situation where everyone will benefit. The particular aims of the different sovereigns created by history and geography might therefore prove to be an advantage for optimal utilisation as compared to what would have taken place had there been one river authority, ascribed the power to act on behalf of all.

To limit reflection on sovereignty and conflict to abstract models regarding principles or legal or conflicting or co-operating relationships between basin states may therefore blur the understanding of underlying issues in a particular river basin and can also hinder a peaceful utilisation of the water course. For a couple of decades, there has been a big debate on whether international river basins will be a source of war or of co-operation between riparian states. Water war theories suggest that, as each riparian state maximises its use of the scarce water resources, conflict ensues and, particularly in water-stressed basins, war may be the end result. In reaction to the water war theories other researchers have advanced water as a pathway to peace theories, suggesting that because of greater interdependence between riparian states they will commonly come together for the core purpose of managing water jointly.

Both these basically deterministic theories can be falsified, and the configuration of power and history and relations among actors in river basins are more diverse than the theories allow for. The society-water interactions are bi-directional, since the social attributes of the actors, their values and interests, and the power relations that influence how the physical environment is conceived, are so diverse. Water's presence within the territory of a nation-state is often very specific to the geographical features of that country, and the people living there will often identify strongly with these water resources and geographical features, considering them part of their national heritage and identity. The place an international water body will have in a nation's cultural life will vary over time, often according to the transaction situation the country is in regarding this water. This fact supports the argument that nation-states cannot be entrusted with the burden of protecting other people's right to the same water. But additionally, societies cannot manipulate their environment at will, since geographical and hydrological factors define what is possible with different means. Thus societies' exploitation of water resources is not only solely based on political, social, economic and technical capacities but must also be suited to the ecological contexts in which such an exploitation takes place. Moreover, as the physical environment changes by natural and human-induced forces, societies have constantly to modify their relationship to the physical environment in order to sustain themselves. These dynamic society-water interactions vary from one basin state to another, particularly in large basins with different ecological conditions. It is these patterns and histories of interactions between the different basin states and their physical environment that influence how these states enter the international hydro-political arena, the strategic choices adopted and the forms of co-operation that are preferred. In river basins it is too easy to conclude that the modern sovereign state is creating or solving the problem of co-operation or conflict (Tvedt 2010). Instead of resorting to general models and universal principles, it is the particular 'rules of the games' in the particular river basin that should be properly analysed in order to avoid conflict and promote further co-operation. Solving of water conflicts is therefore essentially a negotiation of particular linkages, of which the particular geographical and hydrological linkages are but two.

PROPERTY, SOVEREIGNTY AND HYDROLOGY

Historical studies have made it clear that there is no grand theory of development that can explain and grasp change and continuity in international water law; neither national nor international water law has evolved systematically or naturally according to its own methodology or internal laws (Howarth 2014). Resolution of particular cases, especially man/water relations, has often proved to be the 'tail that wags the dog' of legal principle (Howarth 2014). Water law as found around the world today has aptly been described as 'a patchwork of local customs and regulations, national legislation, regional agreements and global treaties' (Dellapenna and Gupta 2014); such laws are reflecting that water law developed in a highly contextual manner mirroring different political systems, religious traditions and economic activities and relations. Some laws are drawn from Talmudic interpretations or from Islamic law regarding Allah as the legislator; others are influenced by European continental law traditions or common law traditions where the judges do not make the law, they only discover it. According to Article 38 of the Statute of the International Court of Justice the sources of international law are a mix of international conventional law, international customary law, the general principles of law recognised by 'civilised nations' and judicial decisions or international case law and the teachings of the most qualified publicists.

The fundamental reason for different law practices in different sovereign states and in different international river basins is not only that all societies and areas have different political, economic and religious histories, but also that they at all times have had to relate to and distribute the particular water running through or underpinning their societies in some way or another. Since legal norms and traditions can only be understood if reference is made to the attitudes of the human beings who established them, to reconstruct their history requires the understanding of the whole situation as apprehended and conceived by the agents whose acts impacted on law developments.

Here it is argued that in addition to political, cultural and judicial history, it is also necessary to integrate analytically the water body subject to law making, since it forms part of this 'situation' as filtered through the lens of the actors. The legal history of international water will therefore also have to integrate in the analysis non-social issues such as the physical characteristics of different water bodies (aquifers, wells and other specific types of running water and river basins, etc.). In order to understand the history of international water law and sovereignty, one therefore ought to study the general historical context in which these laws developed as well as the particular geographical and hydrological features of the legal objects for which the laws were developed. The point we will make is that geography matters when it comes to understanding the development of international water law and the particular notions of sovereignty dominating in different river basins.[14]

The Danube Convention is a case in point and demonstrates the need for a broad, multidisciplinary approach. It was formalised against the background of a very particular historical-geographical water-society relation in the lower part of the Danube River at the time. It was at the end of the Crimean War. The countries in the region wanted that trade on the Danube, that had been such an important waterway for centuries, should no longer be hampered by narrow national interests. Commerce and shipping were almost stopped by hydrological and geographical features of the river. The filling up with sand of the delta, which was shared by different countries, made commerce and trade almost impossible. Boats could hardly go up the river from the Black Sea and vice versa. The situation was especially bad in the year the treaty was signed, 1856, and the mouth of the Danube was littered with the wrecks of sailing ships and made hazardous by hidden sandbars. By internationalising the river this hydrological and natural problem could be solved in the best interest of all concerned. By co-operation among the river states (Great Britain and Italy were also party to the agreement), the common enemy – the sand

– could more easily be moved. It was in fact only by co-operation and agreement that this particular problem facing them all could be solved. The hydrology of this river acted as a definite push towards international, co-operative agreements.[15] Politically as well as historically this was a golden moment, and the countries grabbed the opportunity. Later in the nineteenth century, a number of new agreements relating to the river were signed, and the jurisdiction and powers of what was called 'the European Commission' on the Danube were established (Kaeckenbeeck 1920: 233).

The situation on the Nile was very different. The use of the river was for irrigation and not for transport, and its hydrology has not acted as a push for co-operation since the basic and fundamental feature of the river is that it runs for 2,000 kilometres through one of the hottest deserts in the world, and through two countries totally dependent on water discharge they do not control since it all comes from upstream. The 1929 agreement was therefore not one aimed at solving common problems, but the outcome of political and diplomatic rivalry between Great Britain and Egypt. For political and diplomatic reasons London exploited the fact that the upstream countries at the time could be considered as having no interest in the river at all because they could rely on another part of the hydrological system: rainfall. London institutionalised a policy of limited sovereignty for the East African territories, in order to establish a form of basin-wide co-operation between the two dominant powers, London and Cairo. The 1929 agreement on the Nile cannot be understood without taking into consideration the river as part of a complex and quite fixed water-society system at the time, with three different and interconnecting layers: (a) the river's enormous length, the fact that it traverses extremely different climatic zones, its variable hydrology; (b) technological development and human modifications of the river; and (c) power relations within the Nile basin and especially British imperial Nile strategy and Egyptian ideas about the Nile as an Egyptian river (see chapters on the Nile in this book).

The infamous Harmon Doctrine must also be analysed in connection with the specific ecology of the Rio Grande River and the years of drought that preceded the formulation of the doctrine, as the general applicability of the up to now successful Indus Treaty between Pakistan and India in 1960 has limited universal relevance since the solution of assigning all the water of the eastern tributaries of the Indus to India and the western tributaries and the main channel to Pakistan was made possible by very special territorial and hydrological features that are not found elsewhere in international river basins.

In order to understand the development of international water law it is therefore not sufficient to study the development of law itself; one must also study historical context in a broad sense as well as geography and hydrology. But the importance of geography should not be seen in a narrow, one-dimensional and deterministic way. There are no law-like patterns between geography and international law practices, or river systems and treaty design differences. To argue that the most fundamental elements in the analysis of conflict and co-operation over an international river are the geography of the river itself and the location of each state *vis-à-vis* that river is not helpful. Even in those cases where rivers bind states into a complex web of interdependencies, geography is but only one factor since it is the combined impact of geographical location, economic might, technological capabilities, water management capacities and military muscle that influence symmetry and asymmetry in international river agreements (Dinar 2008: 46). Of course, there are some widespread characteristics. The most important factor of long-term consequences is that bargaining power not available to downstream states may be available to upstream states (Sprout and Sprout 1962: 366). But it is not always the case that whoever controls the upper parts of a river basin has a distinct strategic advantage *vis-à-vis* sovereign downstream actors. In the relationship between Lesotho and South Africa regarding the rivers feeding the urban centres of Johannesburg and Pretoria, Lesotho as the upstream power has become a victim of its location upstream in a river basin controlled politically and economically by a very strong downstream neighbour. Sovereign states with apparently enormous potential water power may turn out to be weak in a given confrontation with seemingly weaker states if analysed from a purely geographical perspective. It has been argued that the geographical position of the state – whether it is located upstream or downstream – is the 'key to this veto symmetry' (Dinar 2008: 45), but there are enough cases from river basins around the world to falsify this general theory. Politics triumph most often – but not always – over geography, at least in the short run.

The popular idea that upstream sovereign states always have a geographical advantage is deterministic, and should be regarded as a dogmatic substitute for concrete investigations. It may or may not be the geopolitical constellation, depending on the geography of politics and economies in a much wider sense than just in relation to the one-factor upstream/downstream dichotomy. The Nile might be a case in point: Ethiopia has been an upstream country on the Nile for thousands of years, but it was technologically very difficult to exploit the river at all there due to a number of geographical factors, while Egypt, located at the river's outlet surrounded by deserts developed as the strongest regional power. Ethiopia was not in a position to exploit its upstream position while Egypt used its downstream position to develop by far the most powerful state actor in the whole basin. And as time passed and the basin entered the Modern Period, Ethiopia was barred from using the little water she could use by asymmetric treaty arrangements benefiting the downstream power. Now this is about to change and any general theory must be able to explain why, until now, upstream location has been a strategic disadvantage. These examples are sufficient to indicate that right and might and location are interlinked in much more multifaceted relations than popular ideas comprehend.

To bring geography into the picture is nothing new. In the 1911 Madrid Declaration of the Institute of International Law it was made clear in its preamble that its principles of law were deducted from 'the permanent physical dependence' of co-basin states.[16] As Bourne summarised it: 'The physical features of a drainage basin, its geography', were now to be 'the foundation of the legal rules applicable to its development.'[17] But as Bourne rightly commented, 'an argument based on geography alone does not carry conviction',[18] due to alterations of river basins by man. To understand the historical developments of notions of sovereignty in international river basins or of international law it is crucial to bring into the analysis both human modifications of the river system and ideas about how the water can best be used and distributed. But additionally, geography is more complex than the Madrid Declaration acknowledged. In both the hydrological and a geomorphological sense drainage basins are dynamic rather than static entities. The processes of fluvial geomorphology shape landforms over and through which the water moves. They influence water table depths and how far water is running underground, they impact on soil profiles and not least on stream channels. One can talk about a 'fluvial hydrosystem' (Petts and Amoros 1996), viewing fluvial systems as interdependent combinations of the aquatic and terrestrial landscapes, as the meandering of alluvial rivers, the changing of river channel patterns, erosional processes and slopes, the change over time of longitudinal stream profiles, and so on. The basin scale, although it is in some cases very large, may nevertheless be too small for the effective study of environmental, economic and political issues. One needs, moreover, to take account of the global nature of the hydrological cycle. The issue of scale has been regarded as one of the major unresolved problems in hydrology (Kalma and Sivalapan 1995; Ward and Robinson 2000: 346), since macro-, meso- and micro-scale are all relative terms.

CONCLUSION

By using water as an entry point this chapter has thrown new light on how to understand sovereignty and the history of the doctrine's status.

It has shown that the dominant interpretation of both Westphalia and the 'death of Westphalia' are based on a neglect of empirical data and a disregard for the particular character of the ecology and economy of rivers. Westphalia was not the birthplace of unlimited sovereignty since it also encouraged and codified co-operation among state actors to improve co-operation on the major continental rivers. The notion that the idea and status of sovereignty are currently and unavoidably undermined by ecology and ecological concerns has moreover been questioned, by bringing forth empirical examples showing contradicting historical developments in some important river basins. The chapter has also shown that although treaty making cannot be understood properly unless being analysed in an inclusive geographical perspective, there is definitely no one-factor causal relationship between geographic position in a river basin and bargaining power. The relationship is far more bi-directional and complex. A critical analysis of the interconnectedness between state sovereignty, history of international law, and water, is important because it will reduce the possibility of self-delusion about progress achieved in theories, laws and practices of international conduct in international river basins.

7

WATER AND INTERNATIONAL LAW

Probably no society has existed without water laws of some sort. The fundamental reason for this unique situation is of course that water is the only resource that all societies at all times have had to control, relate to and often share. But since water, at the same time, always in different ways is running through the societies it helps to create and sustain, and these societies, when exploiting such a resource, enter into a particular relationship with their water systems, different water law traditions have developed in different localities, regions and river basins. Comparative studies of water law from this perspective are an undeveloped field.

In order therefore to understand the great variety in water law systems and their characteristics, it is necessary also to analyse the physical water systems as well as the particular histories of the regions in which these legal traditions were developed. It is crucial to understand that, for a long time, water law developed in a highly local manner that reflected the history, geography and political systems of the areas concerned, and how these contexts of time and specific localities shaped the legal discourse. But, at the same time, it is striking how the different water law systems of the world exhibit certain recurring patterns. This is partly the result of the diffusion or migration of ideas about water management and water law, but it also reflects the fact that water is not only particular, it is at the same time always universal in the sense that water has been the same everywhere: constantly in flux, seeking at all times a lower point, and ultimately escaping efforts at controlling it. This chapter will reconstruct the historical and geographical context of the Nile Waters Agreement of 1929 as a case in point. One of the first places where water law developed was along the River Nile – already in the time of the Pharaohs. The discussions in this part will not focus on this early period, however, but on the development of Nile agreements in the modern epoch, especially during the colonial period when the British Nile empire was the dominant power in the region.

In addition to presenting a historical background to the Nile Waters Agreement of 1929 – an agreement that is still at the centre of the current debate among the Nile basin states on how the Nile waters should be managed and allocated – the chapter will discuss how river hydrology and river physics impacted on the agreement in ways that often tend to be overlooked in legal discourses on river agreements and water laws. Due to the fact that the river systems have helped to create different man–environment relations and development patterns along the long stretches of a particular river (in this case the Nile), the legality or continued validity of agreements concluded at a certain point in time will certainly be questioned somewhere down the time-line. Any accord on the use and allocation of large rivers will, of course, reflect existing power hierarchies in the basin and dominant conceptions of the river system. The problem is that often areas and states along a major river basin develop unequally, and therefore develop uneven patterns of water demand and consumption; this subsequently results in the acquisition and formulation of different conceptions of entitlement and attributes of the river itself.

Long and complex international river systems will, due to different ecosystems or river landscapes, encourage different types of social and economic development along the rivers' banks and tributaries, and hence influence or frame localised use of water over time; often, there is a structural relationship between a particular river basin, its hydrology and geography, on the one hand, and the patterns of 'established uses and rights' to the water in the same river basin, on the other. In this context, too, the Nile shall be a case in point.[1]

Lastly, the Nile Waters Agreement can also demonstrate that co-operation over international river basins will not always, contrary to common belief, erode state sovereignty, but might strengthen it, because it provides an excellent arena for exercising and acquiring state authority. A study of the 1929 agreement may throw new light on the somewhat ahistorical legal debate about the relationship between sovereignty and water law.

THE 1929 EXCHANGE OF NOTES

The Nile Waters Agreement, consisting of the exchange of notes in May 1929 between the British High Commissioner in Egypt, Lord Lloyd, and the Egyptian Government, came to have a profound impact not only on Anglo-Egyptian relations and relations between Egypt and the Sudan, but also on economic developments in Uganda, the southern Sudan and, indirectly, on Ethiopia, up to the present day. Without doubt, it has been an important moment in the history of Nile politics, international river basin management in general and in the evolution of international watercourses law. As an agreement on the use of international river waters for purposes other than navigation, and particularly in presenting a detailed water allocation regime between Egypt and Sudan, the treaty has been hailed as one of the first of its kind in the world.

On 7 May 1929, in one of the letters exchanged with the Egyptian Government, the British High Commissioner in Egypt, Lord Lloyd, emphasised that Great Britain committed herself to guaranteeing Egypt her future water supply. Lloyd wrote that the British Government regarded the safeguarding of those rights as a 'fundamental principle' of British policy, which would be observed at 'all times and under all conditions'.[2] London also accepted the judicial principle that the first user (the word 'first' being interpreted in the historical rather than in the geographical sense) of waters of the stream, i.e., Egypt in this case, should have a priority in the disposal of waters it had hitherto utilised. The treaty made it possible for Egypt to build in the Sudan and other upstream countries water control works necessary to herself, block irrigation works that could harm the Nile discharge in Egypt, and reassert historical rights to waters of the river acquired through long use.

An intriguing aspect of the agreement was that the exchange of letters did not define water rights in quantitative terms. It was, however, accompanied by a technical report of the 1920 Nile Projects Commission which has been interpreted as according Egypt and Sudan 4 and 48 billion cubic metres per year of Nile waters respectively.

The 1929 Nile Waters Agreement was a treaty between two consenting states who wished to regulate their relationship on certain matters; it bound those who signed it, as well as other states, on 'whose behalf Great Britain assumed an undertaking', on the basis of its colonial position, that is, the colonies of Sudan itself, Uganda, Kenya and Tanzania. Whatever regime the agreement formed, it applied only to parties to the treaty as such, and no more. In a legal sense, it can be described as a 'law' between those parties, although it does not constitute a corpus of 'public international law' as such.

What was the historical background and the hydrological context of the agreement on Nile waters, and why is it fruitful to analyse both the particular geopolitical situation and character of the geographical structure of the resource which had been the subject of the agreement?

THE COLLAPSE OF A NILE BASIN REGIME UNDER ONE RULE

By the late 1920s, the pioneers of British basin-wide Nile policies, Lord Cromer and William Garstin, Lord Kitchener and William Willcocks, Sir Wingate and Murdock MacDonald, had all left the scene, and the heyday of British Nile control was already a thing of the past. In 1908, Cromer had confidently declared that 'the Englishman' had taken the entire Nile in hand. For the first and only time in the Nile's history, one might talk of a 'King of the Nile waters'– Lord Cromer. At the time, his and his Government's plans for taming the entire river were very ambitious in comparison with most other river systems in the world, and all the projected dams and water infrastructure installations were designed to serve the overall interests of one imperial authority.[3] But 10 years later, the British hydro-political grip on the River Nile had loosened.

London's policy had always aimed at stability in Egypt and development of the country's irrigation infrastructure, especially related to cotton production during the summer time, or the Nile's low season. The idea was again formulated by Lord Cromer: it was Nile control that should convince the 'oriental mind' that they should accept the West's and Britain's leadership. London's main Nile strategy was that the White Nile, which provided almost all of the water during the summer period, was the most important river at the time and should be used by Egypt. The Blue Nile[4] could not be dammed and the flood water could not be stored for the summer season due to its high concentration of silt. Therefore, construction of the first Aswan Dam, completed in 1902, and the crowning achievement of the Cromer–Garstin regime, was built only for seasonal storage of a relatively silt-free water from the tail of the flood. During winter seasons, Sudan's Gezira area on the island between the Blue and the White Niles would take water by gravity from the Blue Nile after the building of the Sennar Dam; at this time of the year, Egypt did not need the waters of the Blue Nile. These hydrological and topographical facts shaped British Nile policies.

The Egyptian revolution of 1919 set in motion political forces that tore the imperial Nile strategy apart, but did not change London's analysis of the role of the river and the relative importance of its two tributaries for Great Britain's overall policy objectives. The political issue of who should control the use of the entire river system came to play an important but neglected role in the struggle for Egyptian independence. The Nile question became part of the nationalist political agenda. The revolution in 1919 and the British declaration of Egyptian independence in 1922 suddenly changed the political landscape and the context of British Nile planning.

Still, Britain's main strategic aims in the Nile valley remained the same: to secure their political and military position at Suez and to increase the export of long-staple cotton to Lancashire. Achievement of the two objectives was seen as being dependent upon the same factor – increased exploitation and control of the Nile waters. However, the strategy that had been laid down so forcefully at the beginning of the century could no longer be implemented in the 1920s.

Egypt had won formal independence in 1922, but it had a vulnerable geopolitical position as a downstream state, a concern disclosed and continually articulated by the nationalist elite. The Egyptian nationalists sought control of the Nile and regarded Sudan as an integral part of Egypt, but gradually they realised that Britain's policy in the Sudan had effectively weakened Egypt's position there. Britain had 'lost' Egypt, but was still a strong upstream power on the Nile since it had occupied the whole stretch of the river from Aswan to the Great Lakes in Uganda, and worked on strengthening its position on the Blue Nile in Ethiopia. Britain was looking for a means of maintaining its influence and military presence in a country that had opted for independence and where the opposition to Britain was very strong. What options did London have?

BRITAIN'S USE OF THE NILE AS A GEOPOLITICAL INSTRUMENT

London had both the financial and technological capacity to control, or threaten to control, the water discharges of Egypt's real lifeline because of its hold on many of the upstream countries. The following quote from archives of the British Foreign Office demonstrates one of the many secret documents outlining identical visions of Nile control as a geopolitical instrument:

> His Majesty's Government are indeed in the position of being able to threaten Egypt with the reduction of her water supply, and this is sufficient in itself to create a feeling of anxiety and resentment in Egyptians; on the other hand His Majesty's Government cannot offer to increase the water supply of Egypt unless the construction of the Tsana reservoir is undertaken. Once this work is completed, they will be able, without in any way abandoning their power to damage Egypt by reducing the supply, to tranquillise Egyptian anxiety by offering to increase that supply to a very great extent.[5]

The analysis was based upon two elements: the fact that Egypt was a hydraulic state and the particularity of the Egyptian waterscape – there was almost no rain in Egypt. In the southern parts of the country, precipitation could be less than 10 millimetres per year and, in Cairo, the yearly average was around 20 millimetres. Nearly the whole Egyptian population lived on the banks of the river, and all economic activity depended upon it. London's aims at the time concentrated on developing irrigation and cotton production in the Sudan and on encouraging development in the Sudan that would weaken Egypt's position and strengthen the hand of London. Britain regarded control of the Sudan as a means to control Egypt and the Suez, and the diplomatic strategy was clearly formulated in numerous secret Foreign Office memoranda: '*The power which holds the Soudan has Egypt at its mercy, and through Egypt can dominate the Suez Canal.*'[6]

Britain wanted to use its control of the Nile as a means of developing a distinct Sudanese identity *vis-à-vis* Egypt. Hence, water withdrawal in the Sudan (and the plans for the Lake Tana reservoir) became keystones in London's efforts to maintain its regional political influence. Developments in long-staple cotton production in Egypt and changes in the international cotton market made the Gezira scheme even more important to the British industries and also to Sudanese finances.

Since the river runs through the Sudan and it is possible, topographically and geologically, to build large dams on the river and divert the waters into the Sudanese desert, London understood the immense political and economic potentials of Nile control. Quite early on they developed the idea that Sudan could become a new hydraulic state on the Nile, and they also knew that any hydraulic design there would create tensions between Egypt and the Sudan.

The so-called Allenby ultimatum exemplifies this, and should be accorded appropriate emphasis in any broad analysis of the 1929 agreement and its causes. In the mid-1920s, with great fanfare but with little success, the British exploited their upstream control of the Nile as a weapon against Egyptian nationalists. In historical annals, this move has been called the Allenby ultimatum, named after the British leader of Egypt and the Sudan at the time.

While the British work on the Sennar Dam and on the Gezira scheme in Sudan went on, the radicalisation of the Egyptian people continued. The enforced compromise on the upper limit of how much water the new Gezira scheme should be allowed to take did not help much to weaken the nationalist movement in Egypt, while at the same time the cotton industry, both in the Sudan and in Great Britain, thought it a highly unwelcome straitjacket. Since 1912, the latter had publicly referred to the scheme's enormous potential, which became obvious to everybody when work started. The higher cost of the project also encouraged higher productivity goals. Water and cotton were still in short supply, with consequences for corporate profits and the local population. For example, the Government had instructed that all the cotton should be sold abroad; local women were forbidden even to hand-spin cotton. If a man on a pumping station kept back a bit of cotton for spinning, it was regarded as stealing and punishable by imprisonment. According to a British administrator, Sudanese women said: 'Isn't it our land? Why shouldn't we women have a bit of cotton? Truly this Government is hard on women' (Crowfort 1924: 86). The problem was how to get more water to the land.

Meanwhile, in Egypt, the upper classes feared more and more that the agitation of the nationalists had unleashed among the population a political attitude that could also threaten their own position. To dampen this radicalisation, former allies of the nationalist leader, Saad Zaghlul Pasha, were now willing to work with the British. The Liberal Constitutionalists' Party was formed and a constitution was promulgated. In the intervening time, the Makwar Dam was being implemented, then regarded in Egypt as a *fait accompli*. The Wafd won a sweeping victory in the elections and, in January 1924, Zaghlul became Prime Minister. During that year, a number of British officials and Egyptian collaborationists were murdered by hardline nationalists. Then, on 19 November, Lee Stack, Governor-General of the Sudan and British Commander-in-Chief of the Egyptian Army, was assassinated. The assassination was a blow to the Egyptians who wanted to normalise relations with Britain and a debacle for British security in the region; however, it also created an opportunity for tough action.

His Majesty's Government, Allenby and the Sudan Government described the murder not simply as the work of extremists, but rather as the natural outcome of a campaign of hatred mounted by Zaghlul and other mainstream nationalists. A situation had emerged in which the British thought they could clamp down harshly on Egyptian opposition, with some support at home and abroad, and so they did – immediately and severely. First, they implemented the scheme for the elimination of Egyptian personnel in the Sudan, which had been secretly drawn up in 1920[7] in the hope of stopping Egypt being a partner in the running of the Sudan (Vatikiotis 1991: 388).

But what shocked the Egyptians most was the issuance of the Nile ultimatum. As a direct and explicit reaction to the assassination of Stack, the British representative in Egypt, Lord Allenby, on the day of Stack's funeral, went to Zaghlul's official residence making a point of not saluting either on entering or leaving the residence; and while trumpeters played the British national hymn outside, he read out loud his famous Nile notice – on 22 November 1924: 'the Sudan Government will increase the area to be irrigated in the Gezira from 300,000 feddans to an unlimited figure as need may require.' What the Egyptians feared had come to pass.

The British reactivated their downstream complex. London gave Egyptians a demonstration in waterpower that would never be forgotten, and which affected the way British Nile policies were later conceived and interpreted. The area of cotton farming in the Gezira was to be increased without reference to Egypt (the British did not threaten unlimited irrigation in the Sudan, only in the Gezira), thus annulling the commitment made in 1920. Allenby later wrote that his intentions were to impress upon Egypt 'the extent of a Power which the country, to its own detriment and ours, had been too long purposely taught to despise'.[8] This extent of power was the authority to dam the Nile, and he knew that he struck at the very heart of the Egyptian downstream complex. Now the time had come to show a fist, he thought.

A number of important political changes followed. Zaghlul did not accept Allenby's demands and resigned the day after. Ahmad Ziwar Pasha formed a new government, which accepted the British demands unconditionally. At the same time as the British were demonstrating the power of the Nile weapon, they were attacking other Egyptian positions in the Sudan. All Egyptian army units were expelled from the Sudan, and a new Sudanese Defence Force separate from the Egyptian Army was established. The Sudanese battalion that mutinied in support of the Egyptians was annihilated. On 27 and 28 November 1924, more than 20 people were killed. Four officers who deserted gave themselves up, and three were sentenced to death and shot by a firing squad. The ideas of the League of Sudanese Union, which towards the end of 1922 had sent a letter to Prince Umar Tusun of Egypt, stating that in the Sudan there existed a movement 'the purpose of which is to support the Egyptian people', expressing their belief that 'the Sudan should never be separated from Egypt' and exalting the cause of the 'the Nile Valley from Alexandria to Lake Albert', were clamped down.[9]

In Britain, politicians publicly disagreed about this use of Nile power. Ramsay MacDonald, who had just stepped down as Labour Prime Minister and Foreign Secretary (January to November 1924), criticised the ultimatum. He regretted that the British had now told the Egyptian cultivator that '[we] hold him in the hollow of our hands'. As Prime Minister, Ramsay MacDonald had, on 10 July 1924, delivered a speech in the House of Commons:

> I give my word and the Government guarantee [...] that we are prepared to come to an agreement with Egypt on this subject which Egypt itself will accept as satisfactory. That agreement will be carried out by a proper organisation as to control [this did not materialise], and so on, and under it, all the needs of Egypt will be adequately satisfied. The Egyptian cultivator may rest perfectly content that, as the result of the agreement

which we are prepared to make, the independence of the Sudan will not mean that he is going to enjoy a single pint of water less than if he had it and was himself working it.[10]

The Egyptians had been frightened, MacDonald admitted, but he suggested another course more in line with what he called British traditions. They should not 'take a single gallon of water required for Egypt', but should instead get a joint 'board set up to deal with the whole problem of the Nile water in the Sudan and Egypt [...] and you and we will cooperate to produce peace, happiness and prosperity'.[11]

In the latter half of the 1920s, His Majesty's Government worked hard to improve the Empire's public image in Egypt; it aimed at establishing a system for Nile development that was realistic and expansive, and that was adapted to the new political-strategic situation. London clearly realised that there would be no chance of negotiating a new overall treaty with Egypt on outstanding issues like the Suez Canal, unless the political damage of the Nile ultimatum was repaired. On the other hand, the British strategists had reassessed the policy of Cromer and Garstin, now described as being 'too closely associated with exclusive Egyptian control' of the Nile and partly blamed for reassuring what was called the 'monopolistic attitude' so deeply engrained in Egyptian public opinion. London realised that having 'lost' Egypt as a protectorate, they could no longer implement the basin-wide plans of the past, but the Foreign Office in London tried to maintain the role previously occupied by the Nile regime of Lord Cromer and his close associate, the water planner William Garstin. Their aim was to continue as a kind of 'General Command' of Nile development, but in a very different political atmosphere.

During this period, the Foreign Office in London regarded itself as the natural control centre and think-tank for utilisation of the Nile, and hydraulic calculations as part of the diplomatic arsenal. When Allenby had suggested that Great Britain might consider it expedient to seek from the League of Nations a British mandate for the Nile and its waters as distinct from any territorial question, this reflected the mood, but it was a wholly unrealistic proposal, of course. Sitting at their desks in Whitehall close to the Thames, the policy makers and foreign policy bureaucrats in London conceived of the Nile as a river which Britain had both an interest in and a duty to control. In the 1920s, they not only faced nationalists, kings, emperors and rival European and American powers in the Nile valley, but they also had to balance the interests of British companies, the Colonial Office and public opinion at home and abroad. The very complex imperial political set-up did not make it any easier: Britain had a High Commissioner in Egypt, but the country was formally independent although London had reserved the Nile for itself; Sudan was ruled from the Foreign Office in London, Uganda was under the Colonial Office in London, Kenya was about to become a white settler state, Tanganyika was ruled by a British Governor and Commander in Chief after Britain, under the Treaty of Versailles (1919), had received a League of Nations mandate to administer the territory. In Ethiopia, finally, London had a British representative and the 1902 agreement with Emperor Menelik II that dealt also with Nile utilisation; and in Eritrea, Rwanda, Burundi and Congo, London had agreements with the respective colonial powers guaranteeing that these powers would not build dams on the Nile without British consent.

To counter what the British described as Egypt's monopolistic attitude and at the same time repair the damage caused by the Allenby ultimatum, London came up with different initiatives that could maintain its role as the Master of the River. The Egyptian Prime Minister, Ziwar Ahmad Pasha, who had unconditionally acceded to all British demands when taking power after Zaghlul resigned, complained in 1925 that the Egyptian Government had always maintained that the development of irrigation in the Sudan must in no way be of such a nature as to damage irrigation in Egypt or to prejudice future projects which were crucial to meeting the needs of the country. He felt that 'this principle' had been fully admitted by His Britannic Majesty's Government in the past.[12] In early 1925, he asked Allenby to revoke the instructions in his Nile ultimatum, which had so infuriated and shocked the Egyptian public. This gave London an opportunity to declare a shift in policy.

Allenby replied the same day that the British Government was disposed to direct the Sudan Government 'not to give effect' to the previous instructions mentioned in his ultimatum.[13] The British line was now again to tell Egyptians that only Great Britain could guarantee them the water they needed, a guarantee less trustworthy after the Allenby ultimatum than before, it is true, but on the other hand carrying more political weight, perhaps, since the British had already proved the power of upstream control. London now wanted to be seen as a kind of broker between the more aggressive Nile policy being pursued by the Sudanese Government – led by the British but with support from the Sudanese who wanted to invest in profitable irrigation agriculture, on the one hand, and Egypt on the other.

The British strategists now aimed to convince the Egyptian general public that Egypt would be compensated for water taken at Sennar – and with London's help.

At the same time, it was strategically important to break down what London called the 'monopolistic attitude' of Egyptians to the Nile waters. Since the 1920s, they had been discussing whether to establish some kind of Nile Board or Nile Commission which could bring more actors and more countries on to the Nile scene. Cairo was sceptical, but after repeated initiatives from London and in the aftermath of the Allenby ultimatum, a Nile Water Commission was appointed in 1925. Officially, its purpose was to examine and propose a basis on which irrigation in the Sudan could be carried out with full consideration of the interests of Egypt, and 'without detriment to her natural and historic rights'.[14] It should define, among other matters, what the well-informed *London Times* described in its issue of 27 January 1925 as 'the vested rights of Egypt and of the Sudan'.

The aims of the Commission became far less ambitious due to Egyptian opposition. Originally, it had three members, but the chairman from the Netherlands died in June 1925. The other two members were R. M. MacGregor, the representative of the Sudanese Government, and 'Abd al-Hamid Pasha Sulayman, the Egyptian representative. In February 1926, they produced a final report.[15]

The Commission, weak though it was, represented an important break with the past, and its report reflected the new political map in parts of the basin. It can thus serve as an illustration of the political pedagogy of water reports in transboundary or international river basins. For the first time in the river's long history, a representative of an upstream state (the Sudan Government) discussed Nile waters on an equal footing with Egypt. The Sudan was also given permission to have an Irrigation Department under Khartoum's authority; after all, matters related to the Nile had until that time been undertaken under the supervision and management of the Egyptian Ministry of Public Works. The Commission also formally accepted Sudan's right to withdraw water for the Gezira scheme. The Report concluded on the other hand that Egypt should be 'able to count on receiving all assistance from the administrative authorities in the Sudan in respect of schemes undertaken in the Sudan',[16] and very importantly, it was underlined that the Sudan should accept a limited rate of irrigation development.[17]

In a shorter time perspective, it was significant that the Commission abolished the limitation on the cultivated area in the Gezira and substituted it with a volumetric limitation. There were obvious technical and practical arguments for this since it established a more controllable and flexible system. But this change in how water demands were measured technically also gave Sudan an additional benefit that neither the British nor the Sudanese Government disclosed. MacGregor, the British engineer who was in charge of irrigation and Nile control in the Sudan, knew that the official required water/feddan ratio in Gezira had been grossly inflated by the former boss of Egyptian water and Nile control, Murdoch MacDonald.[18] Thus, more land could be irrigated per cubic metre of water than was officially known. MacGregor calculated that it would be possible to extend the Gezira by about 1 million feddans without extracting more water from the river, which meant that the scheme could be expanded without detriment to the interests of Egypt. Allenby informed the Foreign Secretary of this discovery. The experts disagreed about the number of feddans that could now be watered, but the implication was that the Allenby ultimatum turned out to have been unnecessary from a 'water demand' point of view. His Majesty's Government was subsequently, due to the inflated figures produced by MacDonald, given much more leeway *vis-à-vis* the British cotton industry and the Sudan Government that both sought a bigger and more income-generating Gezira scheme.

Because of this 'mistake' in the past, London could have it both ways now; they could have an enlarged cotton farm in the Sudan while at the same time they had been given a hydropolitical space to repair the political damage done in Egypt. What has been interpreted in the literature as a rapid British 'change of mind' *vis-à-vis* Egypt was, therefore, partly an upshot of other factors altogether; MacDonald's inflated water/feddan ratios, published in 1919 and 1920, turned out to be a great hydro-political advantage in the late 1920s. London could overnight, if it so wanted, almost double the irrigated area in the Sudan without taking more of the Nile waters from Egypt, which Egyptians still considered as theirs.

To the Sudanese Government, it was still crucial that the Commission should make it clear that their figures for Sudan water needs were not to be taken as necessarily representing the maximum quantity the Sudan might take without prejudicing Egyptian interests. Water requirements at national level are difficult to establish anywhere in the world and, in a large, undeveloped country such as the Sudan was in the 1920s, the task was almost impossible to accomplish on scientific grounds. At the end of the 1930s, the British estimated Sudan's requirements at about 6 billion cubic metres, or about 10 per cent of Egypt's requirements, while at the beginning of the 1920s, these same requirements were considered to be fewer than 1 billion cubic metres; today, the Sudanese Government argues that the demands are for about 35 billion cubic metres of water.

Khartoum also argued that the Commission should underline that they had not considered the question of *rights*, but had looked at the position solely from the point of view of proposing practical arrangements which could meet the actual requirements of the two countries over the next few years. Khartoum feared that possible restrictions recommended by the Commission might bind the Sudan for ever to limits of water withdrawal which were not acceptable. His Majesty's Government agreed, and the report was carefully formulated in such a way that both parties could be satisfied for the present. The need to decide between conflicting interests did not arise and was postponed into the future.

The conclusions and recommendations of the Nile Water Commission of 1925 were neither accepted nor rejected by the Egyptian Government.

But London thought that the Nile Water Commission was at least a step in the right direction in a period when most other things were going against them in Egypt. When Allenby left office in May 1925, he was succeeded by George Lloyd and in April 1927, Adli resigned, succeeded by Abd al-Khaliq Tharwat or Sarwat Pasha, a Liberal Constitutionalist. He negotiated a draft treaty with the British Foreign Secretary, but failed to win approval of the Wafd, the nationalist party.

THE AGREEMENT AND EMPIRE

British concern for control of the Suez Canal and the military base constituted a stalemate with Egypt. Great Britain needed an agreement with Egypt that could secure its long-term interests. Therefore, in the midst of George Lloyd's authoritarian efforts to restrict the activities of opposition parties in Egypt, and as London dispatched the British fleet to Alexandria to back up its claim that the British Inspector-General's service as Sirdar of the Egyptian Army should be extended, High Commissioner Lord Lloyd sent a confidential letter to Chamberlain in which he proposed to offer Egypt a Nile settlement that could form the basis of a much wider future settlement between the two countries.[19] Great Britain should confirm to the Egyptians that Egypt, as a result of her physical configuration, must rely to a greater extent than the Sudan on irrigation works, and that she must therefore exercise a preponderant influence on the general development of works designed to store the waters of the Nile. Britain would give the Egyptian Government 'all possible assistance'. In view of the news that the British had helped to spread in Egypt – about the American firm in Addis Ababa and the plans of the Emperor to build a dam at Lake Tana – these assurances, they hoped, would be regarded as important by the Egyptians. Britain should protect Egypt against a potential dam on the Blue Nile. But Lloyd's proposals were also subject to important Nile conditions: the Egyptian Government should 'avail themselves of the opportunities thereby offered', i.e., work together with His Majesty's Government in carrying out 'without unreasonable delay' a development programme on the Nile.[20]

The British intentions regarding the Nile Waters Agreement should not simply be seen as the legal institutionalisation of a stroke of sudden Nile altruism, but rather as a diplomatic tactical move within a difficult and contentious political and hydro-political situation. In the literature, Lloyd's role has in general been characterised as 'champion of the rigid safeguarding of British interest in Egypt' (Vatikiotis 1991: 284). In this case, however, he showed tactical flexibility to secure imperial interests. Lloyd hoped that British goodwill regarding the water question would further what was already considered a positive development in Anglo-Egyptian relations and Egyptian Nile politics. Lloyd in line with this also reported optimistically to his Foreign Secretary that Egypt apparently had concrete plans for implementing the great schemes on the Upper White Nile, on which London had worked since the late 1890s. An Egyptian Public Works Commission, of which Lloyd's man, the British water engineer A. D. Butcher, was a prominent member, had criticised the slowness of progress on the Upper Nile.[21] The Foreign Office thought it therefore possible that the Egyptian Government, before it decided to increase the height of the Aswan Dam, would start work on the Upper Nile.[22] But London was once again disappointed. Egypt went for raising the Aswan Dam rather than developing the White Nile reservoirs in Sudan and Uganda. The Foreign Office noted that this was 'wholly detrimental to British interests', the reason, of course, being that it undermined the strategic asset of British control of the Nile upstream. In spite of this development, the Foreign Secretary in London, Chamberlain, supported Lloyd's diplomatic efforts and wrote that he should 'not relax' in reaching agreement with Egypt on the water issue.[23]

I have above described the Sudanese and the British Nile policy. What about the other upstream areas that Britain controlled? Since the great natural reservoirs of the White Nile were located in Uganda, and Uganda was the place where several of the planned dams should be constructed, that country was by far the most important to British Nile strategy. Lloyd and London took steps to bring the Ugandan Government in line in relation to the 1929 agreement. It was important that Uganda should not publicly protest against British-sponsored water plans upstream or demand compensation from Egypt for the planned dams in their country at this particular moment. London knew that the Colonial Office and its representatives in Kampala were sceptical about a British Nile policy giving Egypt too much power over the river to the detriment of the East African territories. But to London, such public criticism at the time was dangerous and would only help to infuriate the Egyptians. The long-term aim was said to underline the necessity for a 'comprehensive agreement' regarding the construction and operation of works which were not in Egyptian territory, and 'for which the consent of both the Sudanese and Ugandan Governments will be necessary'.[24] Lloyd knew that Egyptians feared what they saw as unjust attempts to make use of Britain's geographical position,[25] and one way to remove this fear was to play down these territories' need for Nile waters.

It was important to London that an agreement should be in place before more control works were carried out. Instead, the Egyptian Government proposed that works could be started before any such agreement was concluded, since the latter arrangements would only increase what Egypt considered her 'established rights'. Egypt wanted to raise the height of the Aswan Dam for a second time without having to discuss water allocation issues with the Sudan, while Britain wanted Egypt to take part in the upstream schemes in some way or another, but only if this co-operation was based on an allocation agreement. London thus had to win over those in Khartoum who regarded such an allocation agreement as premature. For its part, the Ministry of Public Works in Egypt told Lloyd that it could not accept any abdication of the control hitherto appertaining to it in the valley as such, due to public disapproval. To the Ministry, a new Nile Board as proposed by Allenby and London was a bad idea, and also the British Government in the Sudan was sceptical because they feared that it would mean they were being forced to consent to dams on the Nile in the Sudan built for Egyptian purposes only.[26]

In the meantime, the Egyptian political scene changed. Tharwat resigned and Mustafa an-Nahhas (Nahas) Pasha, Zaghlul's successor, became Prime Minister. After his resignation followed the brief interlude of Nahas Pasha's Government, during which time the negotiations did not make much progress. The King dismissed him in June and dissolved Parliament in July. In effect, the constitution was suspended, and Egypt was again governed by royal decree under a Liberal Constitutionalist premier, Muhammad Mahmud Pasha. Now an agreement on the Nile had become more likely.

The negotiations took place against a background of serious water shortages in Egypt and conflicts over its use. The 1928 flow was very low. One example among thousands can be given: in April, Lloyd wrote to Chamberlain about the difficulties a British cotton growing firm, the Aboukir Company, was facing due to water shortages.[27] The shortage was particularly marked in the province of Behera where the company had its lands. The company had explained that at the time of their complaint, there were six working days and 12 days of stoppage. On 30 March, which was the last of the six working days, the manager cabled that no water had arrived within 8 kilometres of the tail of the canals. The land would have to go for at least 30 days without water.[28] The result, it was thought, was that thousands of feddans would have to go out of cultivation. When the High Commissioner was sitting down at Easter time to write a telegram to the Foreign Secretary in London about how water had reached no further than 8 kilometres from the tail of the canals that gave life to the cotton seed in the province of Behera, the importance of breaking the deadlock on an agreement on Nile control was made evident both to London and to the new Egyptian Government, whose legitimacy, as in all previous governments in Egypt, rested on its ability to bring enough water to the fields.

THE EXCHANGE OF NOTES REVISITED

This Commission's report suggested that Egypt should be guaranteed water sufficient to irrigate the maximum acreage cultivated up to that time, 5 million feddans. On that basis, quantitative estimates were derived which gave Egypt acquired rights to 48 billion cubic metres. The other Nile valley countries were left out of the picture. The entire flow of the main Nile was reserved for Egypt during the dry season. Egypt was further guaranteed that no works which might prejudice her interests could be executed on the river or any of its tributaries upstream. After 15 July, the Sudan was entitled to take water for the Gezira scheme up to certain maximum daily rates in order to fill the Sennar reservoir, and to flood the area developed under basin irrigation downstream of Khartoum. Although this increase was a far cry from the maximum demands that could be heard in the Sudan, it was a step in the right direction for Khartoum. The agreement broke what they called Egypt's 'monopolistic' attitude to the Nile waters. London was to facilitate the establishment of waterworks upstream for the benefit of Egypt and the share of the Sudan in the Blue Nile was dependent upon the amount of water Egypt could draw from other tributaries.

The agreement has been characterised as being 'solely for the benefit of Egypt' (Collins 1996: 157). It was obviously and, from one point of view, strongly biased in favour of Egypt, but this assessment ignores the intricacy of Nile diplomacy and regional hydro-politics. To London, it was seen as a necessary stepping-stone towards a new general treaty with Egypt; it was far less Egypt-biased than the water policies of Salisbury, Cromer and Garstin. London succeeded in allocating more water to the Sudan, and most importantly, this was formally acknowledged by Egypt. An overlooked aspect of the agreement was that any extension of large-scale irrigation in either northern Sudan or Egypt was regarded as presupposing the exploitation, conservation or damming of upstream waters. By giving the Sudan a legal role in Nile development, London also hoped to realise its role as the strategic-political key through which it was possible to hold Egypt – at Sudan's mercy. After all, it was only two years earlier that a leading British water expert could still justifiably write that the Sudan Branch's main object was to collect hydrological information and study projects for the increase of the Egyptian water supply, while the Inspector-General of Irrigation in the Sudan with his headquarters at Khartoum was responsible to the Under-Secretary of State at the Ministry of Public Works in Cairo (Tottenham 1926: 21).

The exchange of notes on what has been called in the literature 'the real issue' – a plan for hydrological development of the entire Nile basin – was silent, however. It has therefore been described as a testimony to 'a lost opportunity, a tragedy', and the 1929 agreement's 'limited achievements' are reflected in the scant subsequent enthusiasm for more 'cement and stone for conservancy projects' (Collins 1996: 158). But, at the time, it was unrealistic that the parties should agree to such a plan of reservoirs across the basin – both on technological and economic grounds and especially for political reasons. Britain wanted Egypt to implement projects upstream, while Egypt feared such projects under British actual control and instead prioritised the raising of the height of the Aswan Dam because it was within Egypt's borders. The Egyptian nationalists were definitely not in the mood to join hands with their British foe to develop their life artery, although, in the 1930s, they grudgingly accepted the Jabal Auliyya reservoir. The Tana Dam on the Blue Nile could not be part of an official agreement as it was placed on Ethiopian territory. 'Black Thursday' on Wall Street, just some months after the exchange of notes, made investors less enthusiastic for more cement and stone anywhere in the world.

One long-term impact was that the agreement established the Nile basin and Nile waters *de jure* as being more than Egypt's backyard. A clause declared that in case the Egyptian Government decided to construct any works on the river in the Sudan, they had to agree beforehand with the local authorities on the measures to be taken for the safeguarding of local interests. Sir John Maffey, the new Governor-General, immediately interpreted the agreement to the effect that no waterworks could be undertaken in the Sudan without the Sudanese Government's consent and that such consent must be withheld unless the Sudanese Government was satisfied that the work would be carried out efficiently and in smooth co-operation. Maffey thus thought that the Sudan had been given an *effective veto* on any work, unless arrangements, which in its opinion were adequate, were made to safeguard local interests. The British in Khartoum, Cairo and London secretly discussed this interpretation. The Foreign Office argued that Maffey overestimated the strength of the Sudanese Government, since there was nothing in the agreement that forced the Egyptian Government to seek consent from the Sudanese Government, although in most cases this would be a reasonable interpretation of 'local authorities'.

The Government of Uganda protested and 'expressed uneasiness' as did the Colonial Office, because the agreement deprived Uganda of any right to exploit the Nile waters in the country (the same was the case for Tanzania and Kenya and in some measures, Sudan as well). The Foreign Office understood but accepted that the freedom of Uganda would be 'restricted'.[29] The Government of Uganda hoped the agreement would lapse when the projects described in the Nile Commission's report of 1925 had been implemented (the Jabal Auliya dam and Nag Hammadi barrage).

They were resting their hopes on an illusion that 'any obligations which it entails on the Government of Uganda will thereby be abrogated'.[30] They grudgingly accepted the limitations put on their development in the short run, since they thought it would be renegotiated rather soon.[31] Nobody asked Ethiopia for her opinions at the time, and London insisted that the 1902 exchange of notes was legally binding and still in force.

The Colonial Office in London was very sceptical about the agreement, because it hindered development in Uganda. The Foreign Office had given a verbal assurance to the Colonial Office that the Nile agreement would be effective only until works contemplated in the report of the Nile Commission had been completed,[32] knowing full well that the final sentence of Lord Lloyd's letter of 7 May gave the most positive assurances that the agreement would be observed at all times and under any circumstances. The Foreign Office could not completely go back on what it had told the Colonial Office, and their top Nile bureaucrat, John Murray, subsequently wrote a proposed text to the Governor of Uganda to be sent from the Colonial Office, in which it was underlined that the agreement was meant to be temporary; 'on the completion of the works contemplated in the agreement, it will be possible to re-examine the situation as it then exists, and to take into account any requirements of Uganda and other British territories concerned which may then call for special consideration'.[33]

The Upper Nile region was still conceived of by both parties as a barrel filled with water. Although Egypt was given the lion's share of the Nile, the allocation system formulated in the 1929 agreement was basically in line with overall British strategy. It turned their planning conceptions of the past into a binding diplomatic agreement with important implications for the future: London prioritised the central riverine Sudan over the southern periphery, and its relationship with Egypt over those with the Sudan, Uganda and Ethiopia.

London hoped the exchange of notes on the allocation of the Nile waters would improve the general political atmosphere so that a comprehensive Anglo-Egyptian treaty could be reached, while Britain could continue to have strategic control over the river upstream. Egypt refused to accept any treaty agreement that did not include a broader solution of the Sudan question. Britain hoped that by guaranteeing the flow of the Nile, Egypt could accept the status quo in the Sudan. Egypt saw Sudan had become more and more under the influence of London, while Cairo regarded the Sudan as being under the Egyptian crown. The Nile Waters Agreement, one of the most important basin agreements in the first half of the twentieth century, can therefore be seen, at least partly, as an expression of Britain's weakened position as compared to the years before the Egyptian revolution, and partly as a reflection of the convergence of Egyptian perceptions of the Nile as an Egyptian river and British strategic thinking.

The 1929 agreement for co-operation on the Nile was an important step in a development that ended with the creation of the Sudan as a sovereign state in 1956. The countries of the White Nile and their potential developmental needs for Nile control works were sacrificed on the altar of Egypt and the Sudan, the latter's since Sudan's use of Blue Nile water presupposed that Egypt got the entire White Nile. During the time of the British Nile empire, the really conflicting interests between irrigated agriculture in the Sudan and in Egypt were a context for and exploited by London in order to strengthen those political forces in the Sudan that wanted an independent, sovereign Sudan, against those who wanted the Sudan to be united with Egypt in a Nile Valley state. The ideas and practice of state sovereignty were therefore strengthened by both the physical aspects of the river system and by how it had been managed and conceived during the British era.

THE AGREEMENT, INTERNATIONAL
WATER LAW AND STATE SOVEREIGNTY

The Nile Waters Agreement of 1929 should be seen as the outcome of a complex power play between a colonial power, Britain, and Egypt, a formally sovereign state but restricted in a particular way because explicitly its autonomy did not cover foreign policy and Nile-related themes. The agreement's content was the product of a complex history where geopolitics, regional political issues and a particular river basin's hydrology and potential for river management and river control intervened. It was made politically possible in 1929 due to particular power configurations, and the final agreement bore the stamp of the river itself and the hydrological regime of the two major tributaries. The water-society relations in the upstream White Nile countries (much local rainfall in many places, an undeveloped irrigation sector, mostly rain-fed agriculture, at the time no hydropower capabilities) made it politically acceptable, although problematic.

What has been described as the general historical tension between conceptions of state sovereignty and the development of legal arrangements for co-operation over transboundary water resources was not irrelevant here, but took on a very special sequel. The accord between the two states sharing an international river was not based on a development whereby past positions grounded in assumptions about Westphalian notions of unrestricted sovereignty gave way to positions which recognised the need to limit the sovereign discretion of states on the basis of sovereign equality. On the contrary, the 1929 agreement was a water agreement that, long before the 1997 UN Watercourses Convention, accepted and recognised requirements of transboundary co-operation over international water resources. However, the agreement made Egypt sovereign over the whole river flow of the White Nile system, while at the same time, it also established Sudan as a potential sovereign actor, especially in relation to rights of utilisation of the Blue Nile.

This bilateral co-operative treaty was not accompanied by nor led to the establishment of a 'community of interests' approach normally achieved by some sort of joint institutional machinery. A form of co-operative arrangement seeking to manage a river basin as an integrated economic and ecological unit or to achieve the sharing of benefits deriving from shared waters, was not agreed upon. Such approaches were opposed by one of the parties to the agreement (Egypt), and Britain was not in a position to impose them. The agreement did not lead to supranational regimes of water resources management where policy-making authority would be lodged in basin-wide institutions.

Water remains, it is said, the sovereignty issue par excellence in the sense that co-operation over common goods is said to undermine state sovereignty. But that was not the case here: sovereignty was developed, linked to and encouraged by demands and disagreements about the use of a transboundary river. The British used potential disagreements between the Sudanese elite and Egypt over the Nile waters as a means to establish Sudan as a country independent of Egypt. As a move to weaken Egypt's monopolistic attitude to the waters of the entire Nile, encouraged by early British basin-wide, multipurpose river basin planning when also London was mostly or only concerned with the Nile and its potential utilities in Egypt, the issue of the Sudan's demand for more water, also reflected in the Nile Waters Agreement of 1929 and the accompanying text of the 1920 report, was a way of constructing sovereignty as a political issue in the Sudan.

The history of the British Nile empire presents an empirical example that falsifies theories about the evolution of water law and the relationship between state sovereignty and international river basins. London as the 'commando centre' placed clear limits on the authority of colonial governments to act within their borders.

The actual management of transnational water resources has in general more to do with international politics and power relations than with such technical issues as water use practices, assessment of water needs or international water law. But the issue is not only about interstate relations or general social relations, it is also about nature and the physical characteristics of the individual river basin, a fact that is very often overlooked in discussions about general legal principles and evolutions of international water law. The history of hydro-politics in the Nile basin shows this probably more clearly than in any other international river basin due to the combination of the following factors: the Nile's hydrological character, its role as a geopolitical object for generations, and the unique richness in historical source material.

The tension between the two principles protecting 'historic rights' and 'equitable utilization' is evident, and is further complicated when the hydro-historical contexts for development and vested rights are assessed.

8

~w~

WATER-SOCIETY RELATIONS
AND THE HISTORY OF
THE LONG TERM

I have shown that a water-system approach is a fruitful and powerful analytical framework in understanding fundamental shifts in history; it can help explain the 'Great Divergence' and why Great Britain and the West took the lead in the fundamental transformation called the Industrial Revolution, throw new light on British imperial strategies and the mechanisms behind the partition of Africa, make urban developments from the first cities in the Fertile Crescent thousands of years ago to the modern megapolises more intelligible, highlight historical and theoretical aspects of the idea of sovereignty and the role of Westphalia, and deepen our understanding of fundamental belief systems and rituals in the comparative study of religions. This chapter will show how the approach can be employed in analysing the long-term history and central development junctures of individual countries. For didactic reasons this chapter focuses on Norway and Egypt, because they represent polar opposites when it comes to how water runs through societies and how water has been exploited through time.

I choose these two countries not because water is more important here than elsewhere, but because this kind of comparative and contrasting analysis most clearly can demonstrate empirically how a water perspective may open up for less reductionist and more fruitful explanations and analyses of different countries' development trajectories. On the one hand is Norway, a unique 'El Dorado' of perennial running water and with thousands of lakes and rivers. On the other hand is Egypt, almost 97 per cent desert, with virtually no rainfall of its own but totally dependent upon one river fed by precipitation in far-off upstream countries.

WATER-SOCIETY SYSTEMS AND CATEGORISATION OF AGRICULTURE

Since agricultural development is to a large extent a story about how farmers have tried to improve the fertility of the soil they are tilling, a permanent challenge for all rural societies has of course been the 'issue of water'. Nothing can grow and be harvested without the right amount of water and, importantly, too much of it is just as bad as too little of it. In large parts of the world the main challenge for societies has been to bring the right amount of water to the fields in the growing season, and in many countries the priority of kings and emperors, states and local communities has been to win this struggle. In other parts of the world, and more seldom, the task has been to get rid of excess water and to adapt agricultural practices to the fields' ability for self-drainage. Since dissimilarities in agricultural activities and social organisation that have been created and recreated by variations in this variable are so fundamental, all agricultural societies can be typologised according to these two broad categories.

Major periodisations of agricultural history can fruitfully be reconsidered based on new and thorough analyses of shifts and differences in agro-water relations, since food production represents everywhere a form of adaptation to the local and regional characteristics of the physical and engineered waterscape. Of course, different food systems are also intrinsically interwoven into the social matrix of any society and political, economic, cultural and religious premises influence what is grown and how it is harvested, and also what is culturally accepted as food. But wheat and rice, goats and camels, fish and fowl – they all thrive in different waterscapes according to tolerance to drought and waterlogged soil, to degrees of water salinity and stream velocity, and so on. Precipitation will in many areas be the ultimate source of water for food production, and therefore the seasonal variations of how water runs in the landscape and through agricultural lands – annual rains or floods, the absence or presence of which types of water at what time of the year, discharge curves of rain-fed rivers – are physical premises for food production everywhere. The difference between how a nomad society in semi-arid regions or in deserts is structured in different ways from farming and fishing communities living in wetland areas is very easy to observe, but it is at the same time a clear illustration of a more complex and universal phenomenon of how confluences between water and society frame and structure food and agricultural production. Agro-water relations and agro-water variability therefore tend to structure food producing regimes and changes in the character of these confluences and interactions will therefore have profound historical implications.[1]

NORWEGIAN HISTORY AND THE ISSUE OF DRAINAGE

One of the most central axes around which the reconstruction of Norway's agricultural history should be organised is the permanent struggle of farmers to establish and maintain an optimal water balance in the fields and in the soil under the almost unique watery conditions in the country. In order to understand Norway's agricultural history in the long term and in a global context, one has to study and analyse how early farming was fundamentally affected by the necessity to adapt to the self-draining ability of different types of Norwegian landscapes and soils. The research should follow and try to reconstruct how the farmers gradually learned and improved their ability to get rid of excess water and adapt to the types of seasonality in water's behaviour in Norway (water is here, albeit with great regional variations, regularly changing form from liquid water to snow/ice and *vice versa* at the same time as precipitation is perennial). The combined impact of these natural aspects of the water system – seen here as an objective but varying structure that all agricultural communities in Norway had to interact with or relate to and thus were impacted by – and the actions of the farmers *vis-à-vis* this water in their agricultural practices, had vital implications for the very fabric of agricultural societies and settlement patterns. In spite of the importance of these particular interlinkages between these two layers of the water-society system in Norway, the issue of drainage has received little attention and has tended to be overlooked in Norwegian historical research. The most influential books on Norwegian history and the history of Norwegian agriculture have rarely dealt with it. The myopic perspective that helped to make drainage an issue that has received such scant interest[2] has most likely been influenced by a historiographical tradition in which explicit and implicit comparisons have generally been restricted to neighbouring and in this respect basically similar countries: Sweden, Iceland and Denmark. Compared to the earliest agricultural civilisations in Asia and the Middle East (not least Egyptian agriculture) – where the most important task was always to bring water to the thirsty fields – this is, however, a very dynamic particularity of Norwegian agriculture.

The average annual precipitation in Norway is around 1,500 millimetres, or approximately double the European average. Moreover, only 10 per cent of the precipitation evaporates, whereas in certain countries, like the desert country Egypt and countries in the Mediterranean basin, have a rate of evaporation which exceeds the amount that falls as rain. The tens of thousands of rivers, streams and brooks do not take the precipitation as run-off directly into the sea because they flow from the great accumulation of snow and ice in the higher mountains, and snow on the fields melts in spring time when plants start to grow. This fairly permanent aspect of the character of the water cycle in Norway (it has been more or less constant since the end of the last Ice Age), and the way in which water has run through the settlements and as run-off over the land, has forced the issue of surplus water on agricultural communities. Norwegian farmers could not disregard the following natural law: few useful plants will thrive with their roots in saturated soil and waterlogged land will hinder the conversion of organic matter, root development and uptake of nutrients and increase the risk of soil compaction and surface run-off. Water beyond what is needed to wet the roots to their capillary capacity will hinder growth, and the problem is that moisture is stored in the soil, and more in clay soil (which is very common in Norway) or in soil lying over impervious subsoil than in sandy soil. Oversaturation has thus been a constant problem influencing in many cases more than anything else the productivity of the soil and thus the harvest. The challenge of 'vannsjuk' or 'waterlogged' soil has therefore throughout Norwegian history been acknowledged by farmers, although it has by and large been overlooked by historians.

The very early history of agriculture in Norway should in the perspective of the long term be seen as a development by which the farmers learned how to adapt to and exploit the natural process of drainage and self-drainage. The first agricultural settlements in Norway were typically and out of necessity located where the soil was self-draining, as in the Jæren area, and where moraine soil was dominant, as in some parts of Trøndelag and Østlandet. The broad theory of Norwegian agricultural development that is suggested here is that the history and geographical spread of early pre-modern agriculture in Norway to a large extent is a history fundamentally reflecting adaptations to a landscape with too much water, and of finding and settling close to fields where there was soil that could be cultivated thanks to its self-draining properties. Farmers had to determine, as accurately as possible, to what extent the chosen fields were or could easily become waterlogged, and very seldom did they have to worry about scarcity of water and how that would affect their choice of plants and trees or the land value. This was, it is hyphothesised, a main reason why many of the most powerful and wealthy farmers in early Norwegian history often had their farms located on hillsides, while the poorer farmers were left with the waterlogged, albeit flatter land, closer to the bottom of the river valleys. The fact that the waterscape in Norway is characterised by perennial precipitation (the timing of rainfall and its particular relationship to the constellation of labour, cropping patterns and capital requirements are crucial but neglected aspects of rural development in general), modest evaporation and natural reservoirs in the form of snow during winter meaning that the water during sowing time (in spring) is heavily saturated, has been a fundamental structuralising force in agricultural development. It has also, of course, been a structuralising factor making the history and productivity of Norwegian agriculture very different from the history of agriculture in water-scarce countries and in many other countries with rain-fed agriculture.

From very early times, the issue of excess water has also required a proactive response, primarily the removal of excess water from the farm in one way or another by means of gravity. In the Jæren area drainage has for a long period been very important, and as new soil was turned into agricultural land the farmers also collectively took initiatives to lower lakes and rivers to be able to drain the new fields properly. One of the first big farms in Norway was called Sanner, which means 'sand'. There they had managed not to bring water to dry sand, but to make fertile soil, or 'sand', from waterlogged clay by successful drainage. The challenge in most places was to get rid of excess water and only very seldom did farmers need to bring water artificially to the fields.[3] All over Norway, even in the eastern part of the country where rainfall was less than in the west and south-west, farmers have had to work to get rid of water, because if they did not do this, productivity would soon drop drastically. The size of the furrows they dug was a function of two facts and considerations; the gradient of the slope on which drainage tiles could be laid and the amount of run-off. Ploughs were used to ridge the fields, and thus the water could be drained off between the ridges. Such actions were particularly crucial in areas of heavy clay soil and special ditches were therefore not always needed to remove surface water. Presently, more than 60 per cent of all agricultural land is under modern drainage systems alone.

A problem for historians looking for written evidence or material artefacts is that drainage measures in the distant past tended not to leave permanent traces in the engineered or cultivated landscape. There are many ways of draining land, and only in a few cases do they resemble the drainage methods described in Cato's '*Di agri cultura*', in which he talked about how big they should be and how stones should be used. Due to topography and natural run-off ditches, it is therefore easy to overlook and impossible to reconstruct accurately the drainage patterns of the past (unless these were recorded in a farmer's own journal), because open drains after a few years will have been destroyed and cannot be distinguished from the surrounding cultural landscape, and old sub-soil drainage is extremely difficult to uncover.[4] So the fact that drainage ditches are difficult to find in historical sources or recover in nature or in cultural landscapes does not mean that they were not made.

The primacy of handling the problem of surplus water forms one distinctive category of agricultural civilisation from a water-society system perspective. This perspective also means that it will be important to study how drainage technology and drainage as a social activity differed from place to place and from agricultural society to agricultural society, reflecting local run-off, topographies and soil types as well as the character of the social system, culture and entrepreneurial capacity. The drainage technology that was available and chosen in the diverse agricultural communities had social implications, whether the solution was just to use land that was self-draining, to dig open drains that could most easily deal with surface flows, or to establish sub-soil drainage (this is, of course, the rule in modern agriculture). Open drains were dug using spades and the task was quite easy to organise socially in Norway, not least since farms normally had a natural brook or stream nearby into which the water could be led. This made it possible in Norway for individual farmers to drain their land without the co-operation of neighbouring farmers, although it was in some cases necessary or advisable to get agreement from the downstream neighbours about drainage plans. Slopes, hillsides and mountain valleys greatly facilitated the improvement of otherwise poor drainage conditions.

Due to the way that water ran in the landscape of Norway, drainage activities were as a rule organised by the individual farmer and on a small scale compared to the projects of the fenlands in England or the polders in the Netherlands that reclaimed thousands upon thousands of acres. In Norway the individual farmer could quite easily dig the furrows or bury the necessary tiles because of the topography of most farms. Although drainage required some local co-operative effort when farmers sought to obtain a combined outlet, they were also in these cases modest in scope compared to the complex and much more hierarchical social organisation and management required, for example, by irrigation of dry land in regions with scarce water resources. The water-society nexus in Norway did not require a strong, central administration, but could usually be handled and managed at individual farm level. Solving the water excess problem therefore produced and reinforced some of the institutional and structural characteristics of Norwegian agrarian society; it organised a recurrent re-enactment of relative autonomy on behalf of the farmer.

The specific social character of drainage practices can be further understood by the fact that it was located within a national legal tradition marked by a water law which was built on a very rare principle of private ownership of watercourses and water. In continental law as well as in common law, state or public ownership of at least major watercourses and groundwater was normal. In Norway the principle of private ownership of watercourses was established already in the medieval landscape codes (Motzfeldt 1908; Nordtveit 2014). Norwegian farmers were also to a greater extent than European farmers owners of their own land. Even before the Reformation, individual farmers owned about 40 per cent of

158

all farmland. From the end of the seventeenth century until the end of the eighteenth century, nearly all Norwegian farmers became owners of their land. Individual farms could be very widely distributed throughout the countryside since neither land nor water was in shortage, and since water runs almost everywhere private ownership of water was normal. This made it possible to put new land under the plough as an individual undertaking, and the fact that drainage could also be implemented by individual farms reinforced private ownership of both water and land.

A focus on water-society systems can also from another angle show that the unusual egalitarian structure of the Norwegian agricultural society was not only or simply a product of culture and politics, as it has been suggested conventionally. Let me mention one example from the nineteenth century: when the market economy developed, many of the wealthiest farms located on the self-draining lands lost their previous competitive advantage for two reasons. They were often further away from the new roads and railways that were laid out in the valley bottoms and, even more importantly, new methods of drainage made it possible to increase the productivity on waterlogged farms, and new ploughing technology and general mechanisation made the flatness of the land suddenly an asset. The land that had previously been the poor man's land now became the better farm closer to the market. This twist countered the natural tendency in rural societies for inequality to increase.

EGYPTIAN HISTORY AND THE ISSUE OF WATERING

A permanent challenge in Egyptian agriculture has also definitely been the 'issue of water', but in a directly opposite form. Here the main task has been how to bring scarce water in the right amount and time to the dry-baked desert soil and keep it there. In Upper Egypt, in the area around the Valley of the Kings and the Valley of the Queens and the ancient Thebes, several years could pass between rainfall events. In Cairo, average annual rainfall has long been around 20 millimetres. The arid climate of Egypt is characterised by high evaporation rates (1,500–2,400 millimetres per year), and very little rainfall (5–200 millimetres per year), which leaves the River Nile as more or less the only source of water. Water must be added to the desert soil to provide the basis for building up plants' cellular tissue, to enable evapo-transpiration by plants, to help dissolve nutrients in the soil and transfer them from the absorbing organs to the rest of the growing plants and, finally, to receive accumulated salts from the root zone into the soil drainage system. Without water mixing with the desert sand, none of this will happen, and nothing will grow. From one point of view it looks like a paradox that this desert country was the breadbasket of Rome, and the biggest exporter of wheat in the world, providing cereals to the Byzantine Empire and, after the Arab invasion in 642, to the Islamic caliphates in Baghdad and Damascus. In the late nineteenth century, it was turned into a cotton farm of the British Empire. The explanation for these agricultural 'miracles' has always been the Nile and its annual flood, and the Egyptians' ability to exploit their particular water system in the desert, and especially in the Delta.

This 'irrigation question' is therefore the axis around which Egyptian history has developed.[5] A focus on water-society systems in interpreting the long-term development of Egypt will conceive of the River Nile as a specific, *a priori*, supra-individual structure which framed human action and development efforts in Egypt, but in different ways at different junctures of technological development. The regularity of the variations in the water discharge of the river and the amount of fertile silt that it carried down from the Ethiopian mountain plateau made the Nile Delta a green oasis, and created at the same time the particular seasonality of the Egyptian harvests and seasons, and the particular seasonal rhythm of its agricultural life. For thousands of years, the basic strategy was to adapt to the natural floods of the Nile, extending it by the rather primitive technology of flood or basin irrigation. Every autumn, after the flood had passed its maximum, banks were built to keep the silty water on the fields as long as possible in order to saturate and fertilise the soil. In the beginning, the basins were in general small and few, but as agriculture developed and use of water increased, more and more basins were built.

The Egyptian waterscape was much more productive than the rain-fed agriculture in Norway. It helped to create large agricultural surpluses, the precondition for the development of the Egyptian civilisation and the emergence of a state, of cities, of priests and scholars. Herodotus, the Greek father of history, wrote with envy about the Egyptians after visiting the Delta more than 2,500 years ago: 'It is certain, however, that now they gather in fruit from the earth with less labour than any other men' (Herodotus 1960: 111). The combination of the society's dependence on the flow of the Nile and the river's relative regularity stimulated astronomy and mathematics. The need for this kind of knowledge was obviously there, but a need cannot in itself be a sufficient explanation. It was most likely more important that experience had told the Egyptians that it must be possible in one way or another to explain and anticipate the annual cycles of the river, because the Nile flows had an amazing regularity. Artificial watering did over time encourage the development of a state administration to solve problems of a collective nature in relation to the river and its water. The development of the state was, moreover, made quite easy, since by measuring the flow of the river at the Nilometres the state erected it could decide the level of taxation, and since the Nile was the main communication route the state could quite easily control both trade and the movement of people.

In the great debate about the emergence of state institutions Egypt has figured as an important case, and one question has been whether Egyptian irrigation initially presupposed a strong state or whether it developed in a more bottom-up manner. There seems to be a growing agreement among scholars that the strong state developed after irrigation was established. But to understand the later character of Egyptian society it is more important to analyse how the particular Egyptian waterscape affected the agricultural sector and the society at large and over a long time period. Karl Wittfogel famously suggested that Egypt should be called a 'hydraulic civilisation', underlining the prominent role of government and drawing attention to the agro-managerial and agro-bureaucratic character of the society. It was a social order with its own type of division of labour, necessitating co-operation on a large scale. He argued that what developed was hydraulic despotism; a form of 'total power spells total corruption, total terror, total submission and total loneliness'. But to describe this state power as 'total power' overlooks the potential clout of the Nile itself in weakening or undermining ruling elites and dynasties (dynasties have fallen partly due to the effects of bad Nile years). It also overlooks how important contradictions within the elite about how the all-important water should be controlled and allocated can fragment the central power. Analysing the 'limits to total power' with a water-system approach, and locating it in this way within the particular Egyptian waterscape and the water-society system that evolved, will open up new avenues of research. Wittfogel's use of his book as an argument in a political campaign against what he saw as 'oriental despotism' in Communist China and the Soviet Union made him blow out of proportion his important insights into Egyptian history and at the same time disregard the bounty of the Nile's particular hydrology, which allowed agricultural activities to be carried out without heavy involvement and control from the state.

The building of small basins could be done on a communal basis, but the building of large embankments on the riverbanks required collective effort on a big scale. The irrigation channels required much work, both when it came to digging them and maintaining them, and the more numerous they became, the greater was the need for a power that could distribute water among them. And the more impressive they became in length and complexity, the greater was the need for a state to manage them. The corvée-system of forced labour in Egypt, organised by the state with brutality and firmness, was labour primarily organised to dig and clean irrigation canals; it had a profound impact on Egyptian society. The scarcity of irrigation water especially in the spring and summer season (the *sefi* season) and, importantly, gradually more so as the irrigation system increased in complexity, encouraged a strong state in relation to society, so that the water could be allocated in a way that served the interests of the agricultural elite and the owners of large farms, while at the same time the small farmer received enough to keep him 'quiet'.

The very fabric of the Egyptian agricultural society was structured by its physical waterscape and the Egyptians' efforts to adapt to and control it. The waterscape and the dependency on the Nile determined the seasons and defined cropping patterns; it became possible to grow cotton in summer and maize, rice and sugar in winter and spring; and it framed available and practical technological options and formed the background and justification for the most important agricultural festivals. The irrigated agriculture that developed on the banks of the Nile did not initially presuppose a strong state but, as the system developed, the state could grow in strength due to taxation. Gradually the system became so complex that a strong state was needed to maintain the basins and secure an allocation of water in line with the interests of the most powerful landowners. I have shown that studies in agricultural history can be analysed in an agro-water variability perspective.

WATER SYSTEMS, TECHNOLOGICAL DEVELOPMENTS AND INDUSTRIALISATION PROCESSES

Chapter 2 showed the relevance of studying water-society systems for explaining the development of the great technological gap between Western Europe and England on the one hand and China and India on the other hand in the late eighteenth and early nineteenth centuries. This chapter will suggest that in order to understand the different paths of early industrialisation in such diverse countries as Norway and Egypt it is also necessary to integrate in the analysis the particular interconnections and confluences between water and society. No such research has previously been undertaken, but also in relation to the industrial age it is possible to draw a fundamental distinction between two categories of countries or regions when it comes to their water-society systems: those that before the coming of the steam engine were able to advance technologically due to their ability to harness water power in one form or another and those countries that were not able to do so because they lacked appropriate water sources to exploit. The development of diverse hydraulic designs and technology cannot be understood in isolation from the waterscape it aimed at adapting to or exploiting.

Many scholars have argued that the spread of waterwheel technology was one of the major technological revolutions of the pre-industrial era.[6] The waterwheel was able to convert the energy of free-flowing or falling water into forms of power. Until the mid-nineteenth century, the waterwheel normally consisted of a large, wooden wheel (later, it was made of iron), with a number of blades or buckets arranged on the outside rim forming the driving surface. Waterwheels were used for a number of different production activities: milling flour, grinding wood into pulp for papermaking, hammering wrought iron, heating blast furnaces for smelting metals and crushing and pounding fibre for use in the manufacture of wool cloth. Waterwheels were often fed by water from a mill pond, formed when a flowing stream was dammed, making it possible to channel water to the waterwheel via a mill race or simply a 'race'. The main difficulty regarding the use of waterwheels was of course their dependence on flowing water, which limited where they could be located. It is therefore research on countries' proto-industrialisation and early industrialisation that will benefit from studies on the diffusion of waterwheel technology from a water-society system perspective.

A comparison between Egypt and Norway in the use of waterwheel technology is very telling because it highlights the analytical relevance of focusing on differences in water-society systems by signifying vital aspects of the countries' modernisation process in general. Norway was in a truly unique position: 160,000 lakes and tarns cover almost 5 per cent of the country's surface and form natural reservoirs, ensuring for much of the year a relatively even, silt-free water supply, carried across the land by the great number of rivers and brooks, and falling down thousands upon thousands of waterfalls carrying enormous amounts of energy. Egypt was also in a truly unique position: one river with extreme variations of water discharge from season to season, running with a heavy silt load across a very flat plain and river delta, and having one lake, lying beneath the level of the sea. The point here is again to suggest an analytical approach that can throw new light on some of the big questions in national development trajectories.

NORWAY AND THE USE OF THE ENERGY OF WATER

The tens of thousands of small rivers and streams that ran through rural communities in Norway made it possible for farmers almost all over the country to set up their own mills. In the 1830s, there were between 20,000 and 30,000 mills in the country, a remarkably high number considering the country's population. In Ringebu in the County of Oppland, half of the farms had their own mill, and this place was not unique. This meant that farmers in Norway enjoyed a high degree of control over the production not only of their crops, but also of their own food, and did not have to crowd together around mills owned by the state or the landlords and succumb to the authority of the powerful to get their daily bread. In Norway, milling activities could not develop into an arena of social exclusion or production and reproduction of social hierarchy to the same extent as they did in France and many other parts of Europe, where the rivers suitable for turning millstones were fewer and further between and the mills therefore became easier targets for the control of the few over the many. In Norway, mills were spread all over the country as integral parts of local agricultural communities and around them and their operation new mentalities emerged. People who lived where there were no rivers, or on rivers where the current was insufficient or too strong or carried too much sediment for the waterwheel to withstand it, were, on the other hand, simply excluded from taking part in this technological revolution and the subsequent forms of labour organisation. Moreover, these differences also fundamentally influenced gender relations. In almost all countries, women have been responsible for traditional milling, using various stone tools and muscle power. The invention and spread of the watermills thus greatly relieved women's burden and reduced their workload. In fact, the watermill should be regarded as a very crucial, though overlooked step in their liberation process. In addition, the grinding process became much more efficient, and rural societies became accustomed to the waterwheel and all the things that these new 'machines' could do.

An understanding of the structural importance of the particular Norwegian water-society system will have relevance for a host of historical questions. One reason why Ester Boserup's famous thesis on the relationship between population density and development does not fit Norway very well is Norway's particular waterscape. The growing population, that in other countries had to find work within the traditional agricultural production system, could in Norway move into other rural activities such as milling, timber export and sawing, or into fishing. In Norway and other countries where water had for rather a long time been quite extensively used to turn millwheels, the same type of water-driven technology was also employed for many other purposes, such as iron smelting and glass making. Waterwheels were absolutely crucial in this process, because they powered the bellows needed to create the necessary heat to form the metals and the glass. While this form of proto-industrialisation with the help of the waterwheel was virtually impossible to develop in Egypt, in Norway the waterscape lent itself to technological improvements on many fronts. While the Egyptian waterscape had served the country well in the era of agriculture, the seasonality of this mighty and sluggish silt-laden river crossing a flat desert landscape had now become a comparative drawback. The many all-year and quite small Norwegian rivers, on the other hand, became the working horses of a number of industries starting from the beginning of the sixteenth century, especially iron industries located on rivers and where water powered the wheels that powered the bellows of the blast furnace, which again made it possible to create cast iron. Due to the fact that all these rural activities were linked to water, the character of the Norwegian waterscape made it possible for all these types of economic activities to take place within farming communities,[7] an aspect of Norwegian society that came to be formative in the transformation process from agricultural to industrial society.

The rivers became the key to the production of sawn timber, Norway's most important export industry for more than 300 years, from the sixteenth century and well into the nineteenth century. Rivers could be used as a transport route for felled trees, from deep inside the forests down to the fjords. And, most importantly, at the same time they provided the power that could run sawmills. Both England and the Netherlands needed at the time building materials for houses and boats. Previously, planks and boards had been cut with an axe and shaped using hand-adzes. The new boats and buildings required more refined materials. While other countries lacked either the timber or the rivers that were suitable for driving gate-saws, Norway had plenty of forests and rivers that could quite easily be exploited for this purpose. The first saw driven by water was as far as we know operating in Germany in 1337, and the technology came to Norway around 1500. It was immediately put to use on a comparatively large scale. The mechanism was simple. A transmission from a waterwheel moved the saw up and down. The heavy, primitive saw-blade was dependent on a river with sufficient power to keep the vertical waterwheel (sometimes undershot, other times overshot) moving. Norway had the rivers that made it possible to float the logs to the ports on the fjords and, more importantly, these rivers could also drive the gate-saws and in places close to the ships that carried them to markets in the rest of Europe. Norway was able to become Europe's leading exporter of sawn timber thanks to these properties of its rivers; Norway had a waterscape that Sweden and Finland, with their slow-running rivers on flatter lands and estuaries, lacked. The existence of these rivers was of course not the cause of the success of Norwegian timber exports, but they could not have happened without them.[8]

When the rest of Western Europe underwent a further intensification of the Industrial Revolution by the use of coal, few factors were of greater importance in Norway than 'white coal'. After the discovery of electricity and the invention of the turbine, Norway could industrialise on a large scale and very rapidly thanks to the country's great number of waterfalls that could now be exploited for the new technology. Hundreds of rivers and streams descend from relatively high altitudes, and in order to produce 1 kilowatt per hour of electricity, a cubic metre of water must fall approximately 400 metres. It is no poetic exaggeration to call Norway 'the land of a thousand falls', since there are around 4,000 of them. The latent energy in the water from rain and snowmelt in the highlands is therefore much greater than in most other countries, and due to the fact that precipitation falls as snow for part of the year, the country enjoys natural storage of water easing year-round power generation. Norway's rivers, streams and brooks descend in stages, through many nooks and crannies, from a wealth of large and small lakes, and onwards into an even greater number of streams and rivers, day in and day out, year after year. River systems are, moreover, spread out across the entire country, and run in all directions. There is not one river that dominates, gathering in the water from a single extensive catchment area, as does the Nile in Egypt, creating a similar type of ecology and also a similar framework of economic opportunities. In Norway the great pluralism of river systems and waterscapes enabled a relatively diversified, yet locally based industrial and entrepreneurial development.

Norway could have been the first country to be electrified due to its water power, but still the best illustration of the country's waterscape is the fact that, already by the end of the 1920s, there were about 2,000 power stations scattered about the country, while many other countries had only one or very few. Today, Norway is still one of the biggest producers of hydroelectric power in the world, and it was the rivers and the enormous and quite cheap power they produced that created communal wealth and made Norway attractive for foreign investments in the industrial sector. The special relationship between foreign (first European and later American) capital – that was invested in heavy industry located in Norway due to its cheap hydropower – and the Norwegian state, can be seen as a functional, adaptive response to Norway's comparative advantage in the world market based on its waterscape.

EGYPT'S INDUSTRIALISATION AND THE
QUESTION OF THE ENERGY OF WATER

A number of studies have been carried out on Egypt's history of industrialisation, but none has given any weight to the issue of its waterscape and how this limited available options and hindered the country's development.

The spread and use of waterwheels for different purposes were fundamental ingredients both in proto-industrialisation processes and in the development of the first modern factories in the European textile industry. Egypt had virtually no such waterwheels primarily because it had nowhere to site a wheel that could be powered by water. Along the

Nile there were no places with a sufficient or appropriate fall of water or where mill races could be dug. The only place in the country where one could find a watermill was in the Fayum depression, away from the Nile proper, lying below sea level, but a place where there was a periodic Nile waterfall that could drive a waterwheel (Willcocks 1889).[9] In more recent times, some watermills were also established in connection with the barrages built across the river to even out the flow so as to increase the area under all-year irrigation. With the technology available up to the middle of the nineteenth century it was simply impossible to use water as a general source of power in Egypt. Men or women therefore had to use muscle power and a couple of stones to make flour for their bread. And there was no alternative to muscle power in driving the spinning wheels or heating potential blast furnaces.

This situation naturally had an impact on the economic development of Egypt and also on what has been described as the conservatism of Egyptian village life, because there was no technology that could break the patterns of the past. In Egypt there were no substantial advances in available technology between the coming of the Arab Muslims in the seventh century and the arrival of the French under Napoleon at the end of the eighteenth century. Within the orientalist tradition this is often explained in terms of social variables like Egyptian conservatism and religious fatalism, but this is clearly insufficient as an explanation. The structural constraints imposed by the Nile and the Egyptian waterscape should not be overlooked.

The story of Egypt's drive to industrialise in the nineteenth century is very telling from a water-society perspective. Muhammad Ali, the Albanian soldier who had seized power in Egypt in 1805, tried hard to intervene in Egyptian markets in a drastic attempt to foster industrialisation. He purchased cotton and wheat at low prices and sold them on world markets at much higher prices, replacing tax farming with his own land taxes, and using part of these revenues to finance manufacturing investment. His Government also supplied flax and cotton cheaply to encourage domestic textile manufacturing. Muhammed Ali subsidised industry and used non-tariff barriers to exclude foreign competition from domestic markets. In the literature there is still discussion as to whether or not Muhammed Ali was successful in his drive to industrialise Egypt, and to the extent that he failed, what the reasons were.[10] Most agree that he was adamant that he had a clear goal, but that he did not succeed. The question is why?

Most research on Egyptian industrialisation efforts in the nineteenth century has concentrated only on social factors, or on what can be explained by social variables, such as (a) the high level of capital accumulation (Issawi 1982: 188; Rivlin 1961: 61), and the fact that (b) Egypt had 30 spinning mills for the cotton spinning and weaving sector, each with an average of 15,000 spindles, and employing some 15,000–20,000 people (Batou 1993: 184); and (c) Egypt's textile output was consumed not only domestically, but also exported. Some researchers have noted that the factories and technology were labour-intensive, small-scale operations, and that the machinery was not driven by power sources other than human and animal muscle power (Owen 1981:72). It has been suggested that this mode of production was the result of a chosen policy, being rational because labour was cheap while energy and machinery were expensive; Williamson and Panza 2013). The result was that, by 1834, Egypt was competing with Spain for the fifth highest spindle/population ratio in the world (Batou 1991: 183–4), and local factories were able to drive imports of lower-quality textiles out of Egyptian markets (Rivlin 1961: 197). Egypt imposed tariffs on foreign goods and subsidised domestic industry so that it could compete with European industry. There can be no doubt that the British especially resisted these industrialisation projects, since they wanted Egypt basically to be a cotton farm for the Lancashire textile industry. Britain pressured Egypt to open up its economy to free trade. But what is interesting from the perspective of the water-system approach is that the whole discourse on Egypt and its industrialisation has overlooked a striking comparative characteristic of the country: it had no places where water power could be used to propel its infant textile industry. The physical waterscape encouraged and enabled a social practice in relation to the most important and varying resource in the society that reproduced rather than eroded the economic and social patterns of the past.

The country was entirely dependent on the Nile for all the raw materials for its factories. As a producer of these raw materials the river 'delivered'. But the same river failed Egypt as a power source. Egypt had no history of watermills, or of a waterwheel-based industry, or of technical devices outside the irrigation sector, or a history of proto-industries as in Europe. When the steam engine was introduced, it had a very weak industrious or industrial foundation upon which to develop, but then the Egyptians slowly developed their own industry, liberated from the prohibiting power of the Nile.

In Norway, the particular waterscape helped to create a sequence of social development, as, for example, a milieu in which technological experiments and non-agricultural activities at local level all over the country helped to pave the way for the economic processes of industrialisation and a mind-set for modernisation. Moreover, Egypt lacked sites where it could build hydropower plants in the late nineteenth or early twentieth centuries,[11] during the period when Norway was industrialising and modernising. Neither of the two countries had coal, so their sources of power had to be different from that which was driving the Industrial Revolution in continental Europe and beyond in the second half of the nineteenth century. Egypt's only river could not be harnessed for hydropower due to its hydrological and topographical character and existing and available technology. The Nile was too mighty and too seasonal in its fluctuations to be tamed for power production at the time, and to harness it to produce a regular, all-year power supply that became so crucial as production input in the industrial age was virtually impossible. In stark contrast to the situation in Egypt, there are, for example, more than 20 waterfalls within the city limits of Oslo alone that could be exploited as a source of power in the early part of the nineteenth century, as well as hundreds of waterfalls in the country as a whole that could easily be exploited for running the new turbines of the latter half of the nineteenth and the early twentieth centuries. Furthermore, the widespread conflict of interest between irrigation (requiring water to be stored from the wet season to the dry season) and hydropower production (hydropower generation requires a relatively steady flow of water all year round) was no issue in Norway, due to two factors, mainly: the dominance of drainage in agriculture and the fact that the waterfalls that were most efficient as producers of hydropower were often situated in mountain areas, far from the main centre of agriculture. In Norway, because of this waterscape agriculture and industry could develop side by side, to their mutual benefit.

At the beginning of the twentieth century, Egypt and Norway had by chance two of the biggest dams in the world – the Aswan Dam built by Cromer and the British imperialists, and the Vemork and Møsvatn Dam on the Måna (Moon River) in Norway, initiated by a Norwegian entrepreneur but erected with the aid of European financial capital and technological know-how. Comparison of the two dams for water control provides new insights into the two countries' different paths of development and how these were related to different waterscapes. The Egyptian dam was for irrigation and was built and owned by the state. The Norwegian dam was for hydropower and was built and owned by a private capitalist. The Egyptian dam was of course on the Nile, the only place it could be built, and meant a revolution in Egyptian society since it made possible all-year irrigation in much of the country. The Norwegian dam was on one of the very many remote rivers running in the middle of nowhere, and meant the start of the modern industrial revolution by generating electricity on a large scale (in this case to produce saltpeter or potassium nitrate). The Egyptian dam was built by the imperial state at the height of its power, financed by European capital, and was intended by the British to be a symbol of imperial might and 'Western methods' *vis-à-vis* the 'oriental mind'.[12] The Norwegian dam was built by the country's biggest capitalist and financed by European capital, and soon became one of the most potent and cherished national symbols of what an individual entrepreneur could achieve. From a superficial point of view the 'text' of the two dams symbolises only modernity and water control, but more fundamentally they express and point to different long-term development trajectories framed by different water-society systems.

POLITICAL DEVELOPMENTS AND WATER SYSTEMS: 'DEMOCRATIC' RAIN AND 'AUTHORITARIAN' RIVER

A very important, albeit neglected, factor in understanding Norway's institutional and political history is that the localisation and seasonal patterns of precipitation in Norway can be described as 'democratic'. Rainfall has, of course, not by necessity turned Norway into a democracy, because rainfall in itself does not shape institutional aspects of societies in this direct way. The point here is that in order to comprehend the development path that Norway eventually took from quite an early stage, and how it is sharply distinguishable from other regions or countries, we need also to comprehend the 'windows of opportunities' created by the structural fact that water has always fallen everywhere and on everybody's 'head' or land. We need to penetrate further and deeper than tautological explanations explaining democracy and communal autonomy and power by referring to the strong position of democratic ideals and communal autonomy. The success of such political institutions and normative systems in Norway cannot be properly understood without bringing into the picture the long-term structural impacts of the waterscapes on the fabric of agricultural and industrial societies; an exogenous factor that has permeated ecosystems, technological options and systems and forms of social interaction in a number of ways.

The different water-society relations that actors experience as routine will also be reflected in their acts, often 'behind their back', because these kinds of everyday experiences are usually conceived of as normal, as part of life, so to speak. It is precisely for this reason that they are often overlooked. In Knut Hamsun's Nobel Prize-winning novel, *The Growth of the Soil* (1917), the main character is Isak Sellanrå. Sellanrå is able to turn his back on society, walk into the forested wilderness and single-handedly cultivate new land. This fictional character is not the universal figure literary critics have seen him to be, but a farmer whose social act reflects a way of thinking that rationally can develop only in an environmental context where water is no limiting factor to the location of farms. It was an act that was possible only within a specific waterscape, dominant in north-western Europe or in the Eurasian raincoast states. In areas, however, where access to even limited amounts of water has presupposed extensive co-operation and organisation, laying new land under the plough would always involve bringing in the authority of a strong state or leader or some sort of collective, communal organisation. In such a waterscape an Isak Sellanrå would either perish or be forced back into the fold, while in Norway, where Hamsun set his novel, an individual farmer could start on his own and make the soil yield its bounty.

In Norway neither feudal lords, local kings nor the state could monopolise water sources and thus exert vital power over the farmers. The nature of precipitation and the river systems made it possible for the Norwegian peasantry to obtain and maintain considerable economic and political independence relative both to the state and the agrarian elite. People could, moreover, cultivate the soil almost wherever they liked, without having to co-operate with others and without having recourse to a complex social organisation in order to do so. Due to the unreliability of the weather, watery soil and an unfavourable topography, agriculture was always difficult, but at the same time it was possible to pursue farming as an individual activity, or to put it differently; individual ownership of farms became possible because in Norway every farmer had virtually full control over all the crucial production inputs, including the water that made growth possible. The distinctive 'ødegarder' ('homesteader' is perhaps the nearest equivalent concept) bears testimony to this notion; farms were often located far from any other village or farm. The question of why the farmers have enjoyed unusual freedom since about 1200 has been raised before; Ernst Sars explained it with political variables and the terrain, which made it difficult for the aristocracy to subdue the farmers.[13] Kåre Lunden explains this unique situation basically using an economic model, focusing on the soil/work quotient.[14] These explanations may both be useful but they both overlook the fundamental issue: the farms could be organised independently of each other since no one had to succumb to a system of allocation of absolutely necessary common water resources, and nobody could control the small farms by simply controlling the amount of water they needed.

The strong communal autonomy and the relatively decentralised nature of Norway's economy and political institutions reflect the waterscape of the country; or to be more precise: Norway could not have developed the way it has, had it not been for this waterscape. Modern factories and heavy industries, often established by foreign capitalists' investments, could be located close to hydropower sources all over the country, because of the geographical spread of waterfalls producing cheap energy. Municipalities all over the country could generate high income and become rich, because they owned hydropower plants. This fact also made the municipalities as institutions and as a group much stronger politically in relation to the central state than was the case in most other countries. It is difficult to overestimate the importance of the fact that in the industrial era of power they owned their own power sources.

This short analysis of Norwegian history from a water-system perspective has also direct relevance for more general efforts to understand social transformation. Given the historical fact that people throughout the world have varied enormously in how they articulated their interests and aspirations, the question of what determines the concrete political means and outcomes that different groups of people actually have available to them has always been at the heart of social science. Different forms of reductionism have influenced efforts at answering these questions, among them the reduction of political means and outcomes to the simple expression of the population's 'interests' or to a vague but all-encompassing 'political culture'. Some of those who objected to these research traditions suggested ways out that have relevance to our discussion of waterscapes and water-society relations and Norwegian political traditions. The political scientists Stein Rokkan and Seymour Martin Lipset turned increasingly to complex historical explanations for why peoples were 'put on' different paths of development. In the influential Rokkan–Lipset theory of political cleavages in Europe,[15] it was argued that the main cleavages can be linked to national and industrial revolutions. According to this theory, national revolutions produced the classic centre–periphery conflict between dominant and peripheral cultures. The Industrial Revolution gave birth to rural-urban and class cleavages. They insisted that the interaction of economic, political, religious and demographic factors grounded major variables in history. They aimed at explaining the political systems of peripheral areas such as Norway as cumulative consequences of their regions' connections to the chief differentiating processes, and in the view of the authors what they called the Industrial Revolution was such a process since it transformed Europe as a whole. They use Norway as an important case, and one problem with their theory is that it overlooks a crucial empirical trait of Norway's industrial geography and contradicts perhaps the most important aspect of Norwegian society in a comparative context; some of the most important industries of the Industrial Revolution in Norway were rural, or located in a predominantly rural setting due to the central pull factor and geography of the power-producing rivers. The Rokkan–Lipset theory was concerned with the impact and role of geography, but a further tracing of the geographical variations identified by their conceptual maps will not yield large intellectual returns. The search for the origins of the political means and outcomes available to different groups of people in different geographical regions at different junctures in history should continue, but water-society relations should be focused, since such relations had fundamental but as yet little known cumulative consequences in the regions' connections to chief differentiating processes.

Just as precipitation patterns and waterscapes must be integrated into an analysis of democratic development in Norway, precipitation patterns (the desert climate) and waterscapes, and the dominant role of the River Nile, must be integrated into a similar analysis of democratic development in Egypt. It should be noted that the water-system approach suggested here has nothing to do with specific theories, as for example the idea that historically a strong and authoritarian state was a prerequisite for early irrigation societies.[16] The approach is a methodology and an analytical frame, and does not imply specific theories about causal relations. In the case of Egypt it will highlight that as the population increased and the economy developed, the state became more and more important as provider of the most crucial resource. In modern times, as the state and its engineers control the amount of water in the Nile and in the irrigation canals at any time, the state has also become the sole actor responsible for the amount of water reaching the farms. The modern state gets both the acclaim and the blame for the water on the farms, as it controls and distributes all the country's water. With the Aswan High Dam, formally inaugurated in 1971, the authorities ultimately decide how much water the Nile brings, and how much water each peasant farmer is allocated. No longer can the rulers blame the gods or Isis or her tears for failing to provide water for the *fellahin*. Now everybody knows as a fact that it is the state and the water engineers who decide not only how much water the individual farmer can have, but if he is to get any at all. By focusing on the relationship between society and water it makes clear that the individual farmer is left with no choice: any farmer who is not part of an organised social system where water is allocated will get no water. The current relationship between the state and the farmer has thus developed over time, encouraged by the role and character of the Nile, the forms the control over the waterscape has taken, and the role of the Nile in the Egyptian national psyche.

In the 'age of power' the ownership of power sources has been of great social significance for the relationship between society and the state. In Norway the waterfalls were locally owned by municipalities and even by private entrepreneurs, while in Egypt the state became the sole owner and manager of the country's most significant waterfall, the power source of the Aswan Dams. In Norway there were 2,000 hydroelectric plants as early as 1929, spread over most of the country. Water and waterfalls were often even privately owned, while in Egypt the water was conceived of as a gift of Allah and managed by the state, and no individual could own a stretch of the Nile. While the waterscape in Norway made it possible to create a society with multiple centres of power and a decentralised industrial landscape, in Egypt the waterscape stimulated a development with one strong centre. The water-system approach makes it possible to analyse this long-term development by focusing on the interconnections of the different analytical layers and the co-evolution of the waters and the societies created political entities with extremely different social structures, giving widely varying options for democratic development and institutional pluralism.

LONG-TERM HISTORY AND THE PRISM OF WATER

In order to understand the widely differing long-term development of Norway and Egypt it has here been argued that it is vital to analyse the character of waterscapes and the inter-relations between water and society. This chapter has suggested that a hydro-historical approach can offer new explanations to central historical issues and raise new questions and suggest hypotheses about issues not yet researched. It has also discussed how such analyses can be carried out, and indicated that the water-system approach requires knowledge of subjects like meteorology, hydrology, sedimentation processes and the potential of dam and mill sites in order to make sense of long-term historical and social development. The empirical analysis illustrates how social facts can and must also be explained by non-social facts and non-social variables, like the (changing) nature of a river or rainfall patterns, or the character of the local, regional or national hydrological cycle. A history that aims at broad, inclusive analyses cannot be confined to the study of written sources or the enactment of past thoughts alone. It shows that 'nature' or the 'water system' should not be relegated to a 'passive' background or arena for human actions in an introductory chapter, because such systems change over time and the same systems play different roles at different times because societies' ways of relating to them differ and change. On the other hand, it strongly objects to the environmental determinism inherent in some of the aspects of Karl Wittfogel's theory of hydraulic society that stated that both too little and too much water leads necessarily to centralised water controls and governmental despotism.

Different development trajectories should not be seen as a mechanical, necessary outcome of the structure of waterscapes, be it the dominant position of the Nile in Egypt or the 'democratic' precipitation in Norway. The actual history of any country should be understood as the result of an active process, accomplished by, and consisting of, the doings of subjects and their interaction with water over many generations. Individual actors dealing with taming or controlling large rivers or waterfalls, like the engineer William Garstin in Egypt and the capitalist Sam Eyde in

Norway, initiate and push through often dramatic changes in the history of countries. Alterations in hydrosocial cycles have often very far-reaching social implications. In some cases the very formative structure of river systems is changed, translated into new structures by human actions. But it is only when located within these structures that entrepreneurial ingenuity or the revolutionary aspect of water engineering can be understood and appreciated. This permanent and enduring relationship creates a process of emergent causation within ever-changing water-society relations – once-optional actions or additions become more or less obligatory or rational for specific actors and, over time, new arenas and frameworks for social actions are continuously created, but always embedded in historically contingent relations between water and society.

9

WATER AND CLIMATE CHANGE

The climate change research story has by and large been a narrative about temperature and carbon dioxide. Al Gore, the 2007 Nobel Peace Prize laureate, summarised in his Nobel lecture this way of looking at things:

> So today, we dumped another 70 million tons of global-warming pollution into the thin shell of atmosphere surrounding our planet, as if it were an open sewer [...] As a result, the earth has a fever. And the fever is rising. The experts have told us it is not a passing affliction that will heal by itself. We asked for a second opinion. And a third. And a fourth. And the consistent conclusion, restated with increasing alarm, is that something basic is wrong. We are what is wrong, and we must make it right. (Gore 2007)

Most scientists seem to agree that the rapid increase of carbon dioxide in the atmosphere impacts climate developments, but this chapter argues that in order to understand the workings of climate and how society impacts it and is impacted by it, one should follow the multiple roles of water throughout nature and society.

It is the combination of two facts that makes it fruitful to follow how water moves through nature and society in order to understand both climate and its impact on societies and the intricate relationship between the two. Water plays a crucial role in the workings of the climate system and changes in climate in societies will first and foremost manifest themselves as changes in the way water runs in the landscape. This chapter will argue that the concepts of parallel and interlocking hydrological and hydrosocial cycles will prove to be useful for analyses of these relationships and processes and how they develop over the long term. These concepts should be regarded as additions to the three-layered water system, highlighting other aspects of relations between water and society.

WATER IN CLIMATE PROCESSES

The suggestion to 'follow the water' will be regarded by some as controversial and criticised for being 'anti-green', perhaps also representing some sort of climate change scepticism. For the 'green movement', putting focus on one element in nature may be seen as contradicting the aim and ideal that research on climate and climate change must study the climate system as a whole and the couplings between its individual components (see, for example, Bridgeman and Oliver 2006: 1). This holistic approach is based on the idea that it is necessary to avoid what has been described as the main problem in climate research: the practice of isolating individual climatic or atmospheric components. The aim of a holistic approach is therefore to understand the interactions of the physical, chemical, biological and human processes that jointly determine the conditions for life on the planet, that is, to carry out analyses within the framework of an integrated 'earth system'. It is argued that this approach has the virtue of recognising the complex interaction between components and links between the great systems of the earth and that it also makes it possible to analyse the ways in which humans affect climate through the socio-economic system (see, for example, Steffen 2001).

Climate is generally defined as the total experience of how weather and the atmosphere behave in a given region over a specified number of years (*Encyclopedia Britannica*). Because this topic is so broad, climate research has, however, in practice given priority to certain elements of the total climate system, for example, the role of the sun, ocean currents or wind, particularly temperature or, lately, one of the greenhouse gases – carbon dioxide. In contrast to climate research asserting that it studies the system as a whole and this system's relation to societies (the latter also often evoked in holistic terms), this chapter argues that this is practically impossible, and instead it favours a more modest but still very ambitious research programme: that of following the movements of water.

A focus on water may be seen by some as bringing the focus away from carbon dioxide and therefore being detrimental to the fight against global warming. But for some decades, developments in climatology argued that water dominated climate and that between 75 and 90 per cent of the greenhouse effect is due to water vapour and clouds. During the post-war years, the orientation of the climatology tradition moved away from parameters such as temperature and relative humidity towards measurements of flux. Climate, it was also gradually realised, could not be understood as a purely atmospheric issue, since it became evident that both its causes and its effects extend deep into the soil and the oceans.

Research became more focused on the transfer and transformation of energy. It was not air temperature *per se* but heat exchange that came to be regarded as essential to an understanding of climate mechanisms. From taking snapshots of atmospheric data, the task became to understand the mechanisms involved in energy and moisture exchange. Various disciplines have now shown that the fluid element of water is *the* vital component of these mechanisms. The current debate on global warming and Al Gore's and IPPC's strong and almost sole focus on carbon dioxide as the only relevant greenhouse gas and on temperature as the most important issue represent a shift away from this tradition.

Here only some of the main factors that support the idea of following the water in climate research will be briefly mentioned.[1] Several useful volumes about how the atmosphere and the oceans transport energy around the globe are already available, and many of these fundamental physical processes need not be discussed here. The seas cover more than 70 per cent of the surface of the planet, and to an average depth of about 4 kilometres (Sarachik 2003: 129). So enormous are these amounts of water, that the oceans contain 80 times the amount of carbon dioxide stored in the atmosphere. About 5 per cent of the planet is covered by ice, or water in frozen form. The remaining 25 per cent of the area of the planet, normally classified as 'dry land', is crossed by rivers, dotted with lakes, populated with plants, trees, animals and humans that are mostly water. Beneath the surface of the earth and the ocean floor are enormous pools of groundwater (the amount of groundwater is estimated at 99 per cent of all fresh water in the world) that historically have only to a limited extent been part of the hydrological cycle but that now, due to a greater amount of pumping, increase the amount of water vapour. In the course of the past few decades, new and enormous aquifers, such as the Guarani in South America, have been discovered, and in the years ahead the world will be astonished by new discoveries of large 'water banks'.

The water that envelopes the earth is constantly on the move, in its own but still partly mysterious ways. On the continental scale, water is moved from the ocean to the land by winds and energy flows to the atmosphere and back to the sea. The circulation of water between the atmosphere and the oceans, of which precipitation is a small but essential part (it is roughly estimated that every year more than 113,000 billion cubic metres of water fall to earth, enough to cover all the continents to a depth of 80 centimetres), is very complex. It also involves continuous alternations in the phases of H_2O, from a liquid, to ice or vapour and back again. This hydrological cycle is a network of interactions between phenomena on scales that range from the millimetres of a leaf to the global, and timescales from seconds to millennia.

Water evaporates on average by about 1 metre a year, but since the oceans are 4,000 metres deep on average, most ocean water does not normally take part in this cycle (the *average* time that water spends in the atmosphere is nine days, an indication of the fluidity of the system, while any given molecule of water may stay for thousands of years in the sea). Oceans are major factors in determining water surface temperatures and fluxes. The surface evaporation-atmospheric moist convection process, for example, is so strong at the Equator that it couples the atmosphere and the ocean into a single complex body that exhibits a continuous irregular joint oscillation known as the Southern Oscillation. In this process, water is the heat engine that powers atmospheric motion, thus continuously redistributing water mass and solar energy around the globe, while sweeping invisibly through the air in the form of water vapour and flowing across the surface of the earth as run-off. Water is thus not merely a passenger on the passing wind. To a great extent, water creates the movement of air or the breeze that transports it, and hence also regulates temperatures.

At any given time, about half the earth's surface is covered by clouds, which play a vital role in the hydrological cycle and radiation balance since they are nothing but masses of water droplets or ice crystals suspended in the atmosphere (Salby 1996). Clouds range in size from a few metres to hundreds of kilometres. Some regions are very cloudy, while others rarely see a cloud at all. Cloud formation, cloud characteristics and clouds' release of precipitation are key aspects of the rotation of water but also of climate in general. Most of the water in the atmosphere is in its vapour phase, converted to droplets or ice crystals depending on temperature and cloud physics. These fall out of the atmosphere as precipitation, and are replenished by evaporation from the seas and by horizontal and vertical transport within the atmosphere (Dickinson 2003: 124).

It is now undisputed: water vapour is by far the most prevalent greenhouse gas. There is about 60 times as much water vapour as CO_2 in the atmosphere. Also the Intergovernmental Panel on Climate Change (IPCC) acknowledges that vapour accounts for somewhere between 60 and 70 per cent of the greenhouse effect. This has made sceptics question that global warming can be caused by increased amounts of CO_2; since water vapour is the most potent greenhouse gas and is created through natural evaporation rather than human activity, current warming trends are nothing to worry about. But this viewpoint is also too simple: it overlooks the reactive nature of water vapour. The amount of vapour the atmosphere can hold is to a large extent a function of temperature, since the warmer the air gets, the more water it is able to collect.

All also agree that water vapour is a very efficient greenhouse gas and more effective at absorbing the thermal radiation from the earth's surface than carbon dioxide. Incoming short-wave radiation from the sun is let in, but water hinders the longer-wave radiation from leaving earth. There is a maximum quantity of water that clouds can hold before they become saturated, and this process increases approximately logarithmically with increasing temperature (Ward and Robinson: 15). Water, clouds, evaporation and precipitation are thus very significant, but at the same time there are also great uncertainties regarding both measurement methods and the data they produce (Goody and Yung 1989; Hartmann 1994; Lindzen 1996; Kiehl and Trenberth 1997; Held and Soden 2000; Trenberth et al. 2009).

Computer modellers of climate change have, however, tended to overlook or reject the role of water vapour. Some include the positive feedback effect, but although crucial, it is not the most important way in which water is involved in the climate system. The changes to the hydrosocial cycle due to the radical increase in anthropogenically produced water vapour as a result of watering and transpiration losses from farming and gardening, from canals and artificial lakes, from the burning of oil and gas, and losses of water pumped from aquifers, also impact the hydrological cycle and thus global distribution of humidity. The fluctuations in the water vapour content can therefore not be reduced simply to temperature fluctuations. Irrigation and large water projects, such as the South to North Diversion Project in China, the Aswan, the Renaissance and Meroe Dams on the Nile, the taming of the Colorado River and the pumping of large underground aquifers affect the humidity not only in the immediate vicinities of the fields, but far away.

The importance of water in regulating climate is now undisputed, and a growing number of studies confirm its centrality. A typical example is a study published in *Nature Geoscience* (19 January 2014), which shows that during the abrupt cooling at the onset of the Younger Dryas period, 12,680 BP, changes in the way water cycled were the main drivers of major environmental changes in Western Europe.[2] The article argues that the intrusion of dry polar air into Western Europe led to the collapse of local ecosystems and resulted in widespread environmental changes. It also suggests that future climate changes will largely be driven by hydrological changes, not only in the monsoon regions, but also in temperate areas such as Western Europe. It has also recently been argued that human modifications of local hydrological cycles have altered the hydrological cycle in far-off places. Due to irrigation and deforestation in Asia, regional changes in water vapour patterns there changed rainfall patterns in Africa (Gordon et al. 2005). The theories of why Ice Ages happen are many and often contradictory, and none are satisfactory. The extraterrestrial theories – that something happened in our solar system or the Milky Way or even on the sun to start an Ice Age – do not concern us here, but the 'terrestrial' theories allocate a central role to water. Although the 'snowball-earth theory' has met strong opposition, most other terrestrial theories do give the hydrological cycle and changes in it an important explanatory role. Moreover, it is widely assumed that the behaviour of it has largely created and affected the glaciation periods during and between the Ice Ages. Temperature, as compared to water, is a way of estimating climate development and change, but it is not in itself a 'force'.

Between 75 and 90 per cent (although pinning down the exact contribution is very difficult if not impossible) of the greenhouse effect is caused by water vapour and clouds. The jury is still out, however, on the respective roles played by water vapour and carbon dioxide in causing global warming, but evidence points to oceans and the water cycle as being stronger than the effects of CO_2. To suggest that carbon dioxide decides the water cycle is like claiming that the tail wags the dog. But no matter what conclusion the scientists draw on this issue, there is general agreement that the water cycle is absolutely crucial in climate processes.

The fundamental role of water in defining and classifying climate in societies is expressed in the fact that most climate classification systems are largely based on the criteria of local and regional or continental differences in the water cycle. Wladimir Von Köppen's pioneering model for climate zones was based on the concept that native vegetation is the best expression of climate, but this in turn was founded to a large extent on average annual and monthly precipitation and the seasonality of precipitation.[3] C. W. Thornthwaite's influential climate classification method monitored the soil water budget using evapotranspiration or the proportion of total precipitation used to nourish vegetation over a certain area.[4] In spite of all this, our knowledge of the workings of the water cycle is still surprisingly defective.

For example: we do not yet know how much rain actually falls on earth (it could be as much as twice as much or as little as half of typical estimates). We do not know the amount of evaporation. We do not yet know how clouds function or the roles of ocean currents. Measurements of amount of and changes in water vapour in the atmosphere are very difficult to make and the data are therefore basically little more than guesswork. The same holds for summary measurements of spatial and temporal variations in patterns of precipitation. Many comparative data sets are unreliable and not very extensive, partly due to changing measurement methods and because measurement programmes in most countries are quite recent. It is also difficult to identify the causes of changes in precipitation, since we do not know much about the role of crucial cloud systems or how and why they change. Even river run-off data are scarce and in many parts of the world erratic, and in most areas systematic registration of river flows started no more than 100 years ago; even now, in some countries many rivers are still not measured systematically.[5] In some of the biggest and therefore also climatically most vital river basins such data are kept as national secrets for reasons of 'national security'.

The main loop of the hydrological cycle, the web of interactions between fresh water and the oceans, has scarcely been studied at all, and we still know little about how fresh water affects ocean currents, temperature and evaporation (cf. the proposal to dam the Strait of Gibraltar to prevent a new Ice Age because studies of the effect of the High Aswan Dam on the Nile suggested that changing the outflow of the Nile had altered the currents in the Mediterranean and the direction of the Gulf Stream in the Atlantic Ocean would change! [Johnson 1997], or the published plan to dam the Baltic in order to send fresh water to what is expected to become a much drier Southern Europe). The difficulty of observing and estimating the interactions between rain, plants and soil moisture is obvious.

A number of different disciplines have studied different aspects of water, but there is no tradition for following the water. Hydrologists have studied the movement of water through the soil and subsoil and in streams and in natural storage sites such as rivers and lakes. Stratigraphers and sedimentologists study the finished products at the site where they are deposited by water, while engineers, hydrologists and geomorphologists are concerned with the source area, the river network or the basin structure. Oceanographers have dealt with oceans, and meteorologists and atmospheric chemists with water in the clouds and in the atmosphere. Hydrometeorologist study the long-term aspects of the hydrological cycle, while plant physiologists investigate the process of transpiration and glaciologists the formation and dissolution of water in the form of ice. Soil physicists have the moisture characteristics of soils as a major concern. In spite of the fact that water is marked by being in constant flux and having no absolute boundaries, research has in general been compartmentalised.

There are still huge gaps in our knowledge of the most basic facts about water. There are no reliable global maps of precipitation and evapotranspiration, or evaporation from the seas, or of aquifers beneath the ocean floor or beneath the surface of the soil. Elaborate models of different spatial or topical aspects of climate and of the hydrological cycle are therefore often mathematical constructs rather than descriptions of the real, complex movement of water from one sphere to another. Data have been accumulating (see, for example, Huntington 2005), but we still know very little about crucial aspects of how this greenhouse gas functions (see Burroughs 2007: 247–56 for an overview), and, by implication, we also know rather little about climate and what precisely drives climate change.

CLIMATE, WATER AND SOCIAL PROCESSES

A focus on water in climate processes is also crucial from another point of view: that of society. I have earlier called the era we are living in 'the age of water insecurity' (Tvedt 2007). The reason is that the uncertainty about how water will behave in the near and long-term future will structure the public discourse about potential futures and fundamentally inform economic decisions and institutional choices, and impact personal world-views.

What constitutes the 'newness' of this new era is not that water in societies is distributed in new ways or that it is possible or rational to talk about a 'novel climate' as some researchers do. The historical facts are clear: fluxes have always characterised water as a natural phenomenon. What is new is the emergence of a *social fact* related to our understanding of climate that is both of fundamental importance and has never been experienced before.[6] For the first time in history, people in general have become aware of the fact that the hydrological cycle has dramatically changed in the past. It is no longer possible to assume that sea levels, glacial cover, river discharges and rainfall patterns are static back-drops to human civilisation. The new uncertainty about the future of humankind is paradoxically resting on discoveries about the history of water in nature, such as the Arctic's tropical pre-history, and the fact that the Sahara after the last Ice Age was a green oasis with large rivers. It is no longer possible to assume that the water landscape might not dramatically change and challenge the survival of most societies.

It is these uncertainties about water in nature and in human societies that now quite suddenly have become a social and political reality and, importantly, they will remain so. This is because it will always be impossible to predict with certainty what the changes in the water cycle will be, at the same time as even small changes in the delicate balance that many societies have established between water and themselves will have far-reaching, often structural consequences. This uncertainty paradox will lead to a situation where major political struggles will be fought not only over the control of water resources but increasingly over projections of the future of water. Differences in ideas about what this future will look like and the most appropriate ways of dealing with it will be a highly divisive issue, and of a new kind.[7] For example, will few people become more powerful than those who manage to locate their projections of the future waterscape on the 'throne' of public discourse? If everybody believes that the water cycle is created by man, societies will be increasingly vulnerable to acopalyctic and fanatic creeds, and this might eventually lead to all sorts of despotism and anarchy.

Since water is absolutely necessary in all societies and plays both a structural and structuring role, changes in the way it moves on global and local scales will have all kinds of consequences for social development (changes in precipitation will impact on everything from urban drainage systems to patterns of tourism, as changes in river discharges and run-off will affect reservoirs, irrigation, fishing, tourism, transport, ecology, oceans, etc.) in all societies. These changes in how water flows in nature and society will also question the climate classifications themselves.

There has been a rapid rise in the number of studies and research programmes with titles like 'climate change' and 'human adaptations', and so on. However, much of this work tends to miss or gloss over a fundamental conceptual problem: the term 'climate change' is too general for use in connection with empirical data or knowledge of social impacts and adaptation. The IPCC defines climate change as: 'Climate change in IPCC usage refers to any change in climate over time, whether due to natural variability or as a result of human activity.' The term thus includes virtually everything, and empirical study of the social impact of 'everything' is not possible. Analyses of the 'impact of climate change' have therefore either tended to be very general and difficult to verify, or have in practice dealt with only certain aspects of climate and their impact without explicitly justifying the actual focus, thereby making attempts to falsify results difficult, and comparative research impossible.[8] Instead of focusing on the social impact of climate change in all its aspects, research should concentrate on only certain features.

Another argument in favour of such more moderate ambitions is that very few people were doing research on such topics before the 1990s, and that the field therefore is in its infancy. The 'climate change' question is both extremely broad and rhetorically very powerful, and such a mix of politicised, imprecise concepts makes it even more crucial to encourage research with findings that can be falsified. A paradox is therefore at work: the broader the focus, the more difficult it becomes to study the interactions between societies and nature or climate. The proposal here is instead that it would be more realistic and more fruitful to study the role of and changes in the hydrological cycle and how these impact on societies, and how social responses again impact on the water cycle.

There are several reasons for narrowing research in this way. All researchers agree that climate change will also because of feedback mechanisms be clearly and directly reflected in the way water flows in nature and societies. This understanding is also reflected in popular perceptions of what changes in climate will mean to societies. People all over the world are ever more often asking questions such as: Are we living in an era that will bring more rain and floods, or more dry spells and droughts? Will the sea level really rise, and if so, how much? Will the glaciers melt and finally disappear because of global warming, or will they remain the same, or will rivers freeze and glaciers even grow because of a new 'Little Ice Age'? There are predictions that the already highly uneven distribution of water might increase, and that some areas will face increasingly severe water stress. Can we expect mass migration from poor dry areas to more water-wealthy areas and an exodus from dry inland areas to the more humid coastal areas? Water managers are realising that they must adapt to an undefined future and can no longer rely on a hydrology based on simple stochastic predictions.

The changing appearance of water in society has already created what are the most fundamental periodisations of world history: the many Ice Ages. When the last Ice Age peaked 18,000 years ago, the mean sea level was about 120 metres lower than it is today, which had a profound impact on settlement patterns and most other social issues.

Some general consequences of changing water are well known. If the form of water changes and more ice melts, the oceans will expand and thus increase coastal erosion and salt water intrusion in coastal aquifers. Densely populated delta areas and estuaries and low-lying megacities will be threatened, and the general migration pattern from inland areas to coastal land, one of the most fundamental historical processes in our time, will be affected. More moisture will be sent into the atmosphere in the tropics, which will generally accelerate fresh water transport to higher altitudes, which in turn will lead to increasing precipitation in many places and to more extreme events related to land-surface hydrology in others, producing both more floods and more droughts. The subsequent changes in soil moisture will have a huge impact on food production, since soil moisture is a crucial factor in agricultural production. The stability of whole food systems may be at risk,[9] and global migration patterns will be influenced on a grand scale. The relationship between snow and rain will change, with immediate and far-reaching economic consequences.[10] Whether precipitation falls as snow or rain fundamentally affects economic activities and social life, especially in the mid- and high-altitude countries of the world. In the subtropics, climate change is likely to lead to reduced rainfall in what are already dry regions. The overall effect will be an intensification of the water cycle that causes more extreme floods and droughts.[11]

Since this book has suggested a long-term view of human history where a very central and universal aspect of it is how human societies have evolved by adapting to and exploiting varying niches created by the co-evolving of different waterscapes and societies, changes in how water runs on the planet and locally will have far-reaching implications. Physical water systems enabled certain types of vegetation, and encouraged certain economic activities and even world-views. Rivers have always impacted on settlement and urbanisation, in some cases providing inland transport routes and access to the sea (almost all the capitals of Europe are located on the banks of rivers). However, the assets of such locations historically can be a liability and in an era of growing uncertainty about the future of water this is especially so. Humans' relationship with water is highly susceptible to damage caused by climate change, due to already established delicate balances and niches.

Contrary to common belief, modern societies have not liberated themselves from the power of water, although modern technology has to a large extent liberated society from the tyrannical, locational power of water. Most people in the world live in a river basin and more than half of the world's population lives in river basins shared by two countries or more. Some of the riparian systems in dry areas play an absolutely crucial role in the life of societies, and they are particularly vulnerable to changes in precipitation patterns due to their high levels of exposure and sensitivity to this type of climate stimulus, while others will display strong resilience. But in general, riparian systems are likely to become what we might call 'adaptation hotspots',[12] as will coastal deltas and cities due to the expected sea level rise. Research on potential impacts of climate change should therefore also for these reasons follow the water.

The implications of changes in the water cycle's impact on society, compared to that of temperature, wind, and so on, can be studied empirically in many locations, since the changes are easily identifiable and measurable (amount and type of precipitation, evapotranspiration, run-off, river discharges, sea level rise, and so on). They can also be studied comparatively, since all societies share social experiences with changing waters. Moreover, different environments, such as wetlands and prairies, snowfields and deserts, river valleys and river deltas, exhibit different regimes of precipitation, evaporation and stream flow, and each therefore presents different challenges, benefits and threats to which societies have adapted and managed in one way or another. Following water also makes it easier to describe and analyse how different societies and parts of society react differently to changes in the climate.

One premise of the argument that it is necessary to focus analyses of the impacts of climate change more narrowly is that the real world of climate and its impacts in society is so complex that it is essential to isolate parts of this reality for analytical and empirical purposes. Another is that this necessary decomposition of climate into more simplified structures for analytical purposes should not be accidental, pragmatic or implicit. This chapter has put forth arguments for why a focus on the hydrological and hydrosocial cycles can be a fruitful decomposition of this broader reality. The ultimate aim of such a research strategy will be to link these simplified structures so as eventually to become able to describe and model the totality of climate and society/climate relations, since it is a truism that the real world is continuous, and isolated structures are always artificial, 'invented' facets of reality. The selected water-system approach should be complex enough to produce a high degree of internal coherence, so that studies will yield significant and useful results. At the same time, the approach is simple enough for comprehension, comparison and empirical research. To study the two notions of water cycles in conjunction addresses multiple concerns and therefore also enables analyses of a hierarchy of an organised, interlocking set of systems.

AN ANALYTICAL APPROACH AND A PAIR OF CONCEPTS

There can be no doubt that dominant social science has basicallly been uninterested in the hydrological cycle as a natural, physical phenomenon of relevance to our understanding of societies and development trajectories. Recently it has been suggested that the concept of the hydrological cycle should be discarded altogether, because it should be seen as a technical, natural science concept; as a concept of the 'hydrologists' or water planners, and on this basis rejected ontologically. The concept should also be discarded on epistemological grounds, since it is argued that the hydrological cycle no longer exists. Humans affect it to the extent that it should rather be renamed. For both these reasons the term hydrological cycle should be substituted by the term 'hydrosocial cycle', it is argued. This book firmly maintains, however, that the term hydrological cycle is very fruitful from a natural and social science point of view. It has proved its usefulness in understanding climate and the behaviour of water as part of the earth system, and in managing water in societies. It retains natural processes as a factor in social development and it acknowledges the role of the hydrosphere and water in the climate system.

The term 'hydrosocial cycle' as used here should not replace the term 'hydrological cycle'; on the contrary, it is in contrast to and in relation with the latter concept that the term 'hydrosocial cycle' acquires its identity, meaning and relevance. The 'hydrosocial cycle' does not imply that the hydrological cycle itself has become 'social' or that it is no longer 'natural'. Rather the use of the two concepts focuses on parallel though interacting processes that in the long run can be seen as a non-dialectical relation where they impact on each other and are impacted on by each other. The hydrological cycle is the fundamental and most powerful cycle in nature it both impacts on and is impacted on by the hydrosocial cycle. The latter focuses on how water travels through societies and establishes a comprehensive framework for analysing its entire route through societies, and also how it impacts on the hydrological cycle over time and in place. It will also enable analytical integration of another social fact: from a social perspective there are many different water cycles in nature, and the hydrological cycle cannot be understood outside of the social. The implications of only talking about the hydrosocial cycle will be that all forms of development will be reduced to outcomes of human action, and human action only, as if the natural workings of the Niger delta and the cloud systems over the North Sea have no social relevance any more or should be reduced to a social fact and variable.

In order to understand how climate and society are impacted on by water and how water (and nature) impact on society, we need to study both the hydrological and hydrosocial cycles and how they interact and relate to the rest of the climate. The movement of water in nature is impacted on by its 'social experience', and it is a truism that the water cycle is always understood through certain social filters. Since societies have developed through interacting with their waters, it is necessary to study both the physical aspects of water as part of the entire climate system and the water landscapes modified and conceived by human beings. This pair of concepts also underlines the fact that the object of study is the same water since its character does not change whether it is observed in the hydrological or hydrosocial cycle, only its role in nature and its relationships to the human world do.

In order to be able to meet the challenges of the future waterscapes, we need to understand past interactions between water and society, not because they can yield any simple solutions for tomorrow – history never does – but because they broaden our understanding of what is possible. Because water is of fundamental importance in all societies, it can be regarded as a kind of talisman for the continuity and deep structures of history and human evolution. All climate changes in the past have been manifested in changes in the water cycle. The migration patterns of hunter-gatherers were affected by where they could find water, and their routes were fundamentally affected by changes in the waterscape, such as the retreating glaciers in Scandinavia and the disappearing rivers of the Sahara. The settled agricultural civilisations on or near riverbanks in Asia and the Middle East were vulnerable to changes in precipitation and river discharge, as the position and spread of the story of the Great Flood testify. For a long time, the collapses of the Indus, Sumerian and Mayan civilisations were simply seen as the outcome of cultural decadence or of foreign invasions and attacks. Now we know that these factors are not sufficient to explain why these civilisations disappeared. Research has established that the fate of the Mayan civilisation was also affected by changes in the in the hydrological cycle; and that recurrent droughts in around 800 weakened the economic base and cultural position of the Mayan leaders who led a rain-based, agricultural civilisation in a seasonal desert environment. We now know that the Indus suffered from bad years and from great floods and that tributaries like the Beas River migrated away from the civilisational centre, Harappa; more recently, it has been argued that the fall of the Chinese dynasties was caused by failure of the monsoon, just as researchers have previously argued that Egyptian dynasties crumbled due to successive bad Nile years.[14] The causal relationships between changes in the water cycle and social impact might have been more complex and less unidirectional than these interpretations suggest, but there is no doubt that changes in climate and the hydrological cycle weakened those civilisations. Everybody who is unaware of the waterscapes and their transformations in the past is doomed to misunderstand the future, and politicians and planners will not manage well the future challenges of changing water cycles if they do not know anything about how societies have adapted to past changes in waterscapes. It is therefore imperative to develop further what is here called a hydro-historical approach.

Research into the history of water-society relations might improve our knowledge about changes in the hydrological and the hydrosocial cycles and also about past interactions between climate and society. Chronicles of religious centres and books of kings and rulers have recorded water-related events due to their importance for societies, and more and more attempts will be made by different natural scientists to reconstruct past river discharge profiles, flood and drought occurrences, severe frosts and heavy snowfalls, employing different sources and methods, from satellite images to pollen analyses. The growing feeling of uncertainty also requires new perspectives, approaches and methods of research on modern society. The genie – our uncertainty about the future waterscape – is out of the bottle and can never be put back. Since we are living in an era marked by these uncertainties about water it will be more important than ever to understand how societies and water interact. The complex roles of water in societies and the links and feedbacks between water in nature and in the human world have become for good far more than an issue of merely academic interest.

10

~w~

A CRITIQUE OF THE SOCIAL SCIENCE TRADITION

In the above empirical section this book has shown the fruitfulness of the water-system approach in throwing new and interesting light on some of the most researched and central questions in social science and history. This section will discuss how bringing water-society interactions into the picture also challenges fundamental theories and concepts about society and social behaviour.

The first part locates the water-system approach and methodology – as proposed in this book – within the dominant research traditions, by discussing and criticising the viewpoints of the founding fathers of social sciences, Émile Durkheim, Karl Marx and Max Weber, and by contrasting them with the ideas of Bruno Latour and Ulrich Beck and the dominant historiography of the last century. Claiming that the most influential research traditions have been basically waterblind gives a historical-conceptual background to the fact that very few studies of social and historical development take as their starting point the knowledge that our planet is the Water Planet and that evolution, development trajectories and social life are written in water. The architects of modern social science disagreed on most issues, but agreed on one thing: that nature could and should basically be left out of analyses of societies, particularly modern societies, arguing in favour of a sharp distinction and dualism between nature and the social.[1] It also assesses influential attempts at dealing with nature-social relations, and shows how problems of nature determinism can be overcome with the help of the water-system approach.

The second part discusses and criticises Anthony Giddens' structuration theory, Fernand Braudel's notions of time, Garrett Hardin's model of the 'tragedy of the commons' and the widespread argument with ontological implications that 'nature is dead', all analysed with a water-system approach and with a focus on water-society relations and interconnections.

Finally, this part suggests that it is both methodologically necessary and analytically fruitful to deconstruct nature as one analytical unity or entity when mapping and analysing nature-society relations and interactions.

THE FOUNDING FATHERS AND THEIR REDUCTIONISM

More than 100 years ago, at the very end of the tumultuous nineteenth century, when the second wave of the Industrial Revolution had just swept over Europe and the USA, when social relations had been changed fundamentally by capitalism and urbanisation, and old world-views had been shattered by the scientific revolution and new theories of evolution, Émile Durkheim (1858–1917), a young French sociologist of Polish descent, was in Paris, writing the preface to Volume I of *L'année sociologique*. He had just founded the journal as a mouthpiece for the sociological movement he sought to further.

In 1895, Durkheim had published his very influential *Les règles de la méthode sociologique*, making it crystal clear that his mission was to establish a new scientific discipline. He had held a chair in social science in Bordeaux since 1887 and, by the end of the century, his thinking about social science was fully formed. He was working tirelessly to strengthen the prestige of sociology. An indication of how much his writings were influenced by his project to convince the French Government and public of the virtues of the new discipline was that he also wrote that France was 'predestined' to play a central role in the discipline's 'formation and progress', because of France's 'native qualities' (Durkheim 1960c: 383).

Soon to become the first French professor of this new discipline, Durkheim acknowledged that as a scientific field it was still in its infancy, and he wrote that in the 1880s there had probably been no more than 10 sociologists in all Europe (Durkheim 1960a: 354). In his introductions to the pioneering journal he laid out the boundaries and justifications for how the new discipline could grow, primarily concerned with how the aims and topics, methods and identity of the new discipline could be validated and delineated. The central task was to define and uphold its autonomy and distinctive character so as to guarantee that it would not be merged with other disciplines with stronger and longer traditions (Durkheim 1960b: 341–53).

Durkheim defined the new discipline in a way that would make social science more relevant and respected with all that would bring with it in the form of support from the Government and new positions at universities: 'The principle underlying this method is the principle that religious, judicial, moral and economic facts must be all be treated in conformance with their nature as social facts' (Durkheim 1960b: 348).

The existence of social science 'can be justified only if there are realities which deserve to be called social and which are not simply aspects of another order of things' (Durkheim 1960a: 354–63).

Durkheim emphasised that the task of the social scientist was to study 'social facts', and that these social facts could and should, whether they were morphological or norms backed by sanctions, be explained in terms of other social facts. His fundamental conceptualisation and justification of social science accepted that the physical environment might be a prerequisite for social life,[2] but the point was to subordinate them to social facts within the new discipline of sociology. The inter-relationships among the various facets of society were conceived of as cohering into a unity – an integrated system with a life of its own, detached from nature and external to the individual. The justification for these viewpoints lay in explicit and implicit concepts of nature or the physical environment. Durkheim very seldom mentioned nature or the physical environment in his works, but the few references he made are interesting and telling. Nature changed so slowly, Durkheim wrote, that it could not explain changes in societies (Durkheim 1984 [1893]: 285–6). Since regularity and monotony were characteristics of nature, and social science was concerned with change and modernity, there was even less reason for social science to concern itself with the physical environment. He did not once discuss the relationship between nature and social variations or waterscapes and social diversity, as if nature and water were identical or so similar that they had no bearing on social organisation. Durkheim also saw nature as a unity; he wrote that 'the course of nature is uniform' and this 'uniformity' was the reason why it did not produce 'strong emotions' (Durkheim [1925 2nd revised edition] 1976: 85) or interesting social facts. He wrote: 'It requires culture and reflection to shake off this yoke of habit and to discover how marvellous this regularity itself is' (84). And he continues: 'Nature is always and everywhere of the same sort. It matters little that it extends to infinity: beyond the extreme limit to which my eyes can reach, it is not different from what it is here' (Durkheim [1925] 1976). The savage 'could be filled with admiration before these marvels', while the modern soul had been 'much too accustomed to it to be greatly surprised by it' (Durkheim [1925] 1976: 85). Based on this sharp distinction and the dichotomy of 'nature' and 'society', or the 'physical environment' and social facts, social science could develop as a separate, autonomous discipline (Durkheim [1904] 1966: 145.) While the natural scientists had natural facts as their object of study, social facts should be the object of study for social scientists. This distinction could safely be drawn – and this aspect of Durkheim's thinking is overlooked by other commentators on Durkheim – because he held to the idea that the physical environment or nature had uniform, basically unchanging characteristics, and did not produce strong emotions in modern society.

Karl Marx's (1818–83) fundamental idea of social evolution regarded development basically as a continuous differentiation of society away from the influence of nature. Marx's basic premise regarding the relationship between nature and society in modern society was that capitalism liberated man from the traditional, localised dependency on nature, and with it the absurd 'nature idolatry' with which this relationship was associated.[3] Nature might be relevant to our understanding of early agrarian societies (although it never influenced his general universal theory of historical development) but in the modern, capitalist world it was not 'nature' that fettered humanity or influenced human societies – it was capitalism. Marxist philosophy and theory of history thereby assessed emphasis on nature to be not only irrelevant but implicitly a thing of the past. A concentration on the various roles of nature in diversifying or impacting patterns of historical development and of contemporary societies was therefore later regarded as a criticism of Marxism and Marx himself, or even worse, as revisionism.[4]

In Marx's works there are, however, comments that show that he himself took a more ambivalent approach to the role of nature and to the way that the distinction between nature and the social sphere should be drawn. In his 'Paris Manuscripts' (1844) Marx argued that human beings are part of nature: 'Nature is man's inorganic body – nature, that is, insofar as it is not itself human body' (Marx 1959 [1844]: 31). That man's physical life is linked to nature means simply that nature is linked to itself, for man is part of nature, he argued in the same manuscript. In *Das Kapital* Marx hinted at the role of nature, especially in the Nile valley at the time when civilisation had emerged (he talked about 'the mild air there' (quoting Diodorus), the papyrus stem and marsh plants, and claimed that children therefore were very easy to bring up in the Nile valley and this minimal cost in reproducing the labour force was a main reason for the Egyptians' ability to build the pyramids![5] Marx also suggested a nature-deterministic explanation of the 'mother-country of capital' and the 'physical basis for the social division of labour', which, by changes in the natural surroundings, 'spur' man on to the multiplication of his wants, his capabilities, his means and modes of labour. But favourable natural conditions alone gave only 'the possibility, never the reality, of surplus labour, nor, therefore, of surplus value and a surplus-product'. Marx was also interested in the exchange and transformation of matter, energy, labour and knowledge between the social system and nature, and in the process of labouring, nature was seen as a kind of substratum. He called this 'social metabolism' and wrote about the potential exploitation of natural resources by capitalist production. Marx's dominant view on nature had, however, some of the same traits as Durkheim; there was a certain kind of antagonism between man and nature and nature was generally seen as an undifferentiated material basis or condition for social life, and progress consisted in the movement from naturally determined human relationships to historically evolved social relationships.[6]

But in Marx's writing there is one striking exception to his general lack of interest in and disregard for geographical structures in general and water in particular. What he called the 'Asiatic mode of production' was explained basically as a product of geography, or of particular waterscape. A brutal ruler controlled a semi-arid environment and a river valley, ran armies of bureaucrats and soldiers, and regulated the way great rivers run and exploited them in large-scale irrigation projects. The state became extremely strong relative to the rest of society, and the despot above and the masses below prevented the emergence of a middle class and a bourgeoisie such as had emerged in Western Europe. One might say it was yet another variant of the old theme: the static East contrasted with the dynamic, progressive West. This theory puzzled later Marxists when they tried to make a coherent development theory of Marx's writings, since in this case he gave nature a decisive role in forming even the most fundamental social and ideological structures not only of one society in Asia, but of Asia as a whole. Marx explained this area's particular social development as a result of the existence of large river basins and how these determined the emergence of specific forms of production that structured society in general. But Marx and especially his followers did not develop this type of analysis further because doing so would presumably have threatened to undermine the Marxist theory of history which was soon regarded and revered as a universal theory of history.

The notion of 'the Asiatic mode of production' was a rather mechanistic model of how waterscapes in Asia directly affected societal formations. Marx here analyses relationships between geography and social development as if the river systems of Asia are so similar that they can produce the same social and economic system. He did not differ between river systems and irrigation systems in China, India and Sri Lanka, all of them being irrigation civilisations, but developed along very different paths. If Marx had been familiar with the system of the irrigation kingdoms in northern Sri Lanka, or with how these compare with the major river basins and monsoon patterns of China (and these are also internally very diverse), and how these again are quite unlike the river basins of the Indo-Gangetic plain and the Deccan Plateau, like the Himalaya and Karakoram ranges, the Vindhya and Satpura ranges in central India and the Sahyadri or Western Ghats in Western India, he could not have developed the simplistic and water-deterministic theory and model of the Asiatic mode of production. Marx's generalisations about the Asiatic mode of production should be seen as representative of an entire trend: basically uninterested in nature and in water, the social scientists tend to advance explanations that are deterministic when natural issues are turned into a variable in social analysis.

Max Weber (1864–1920) was a pioneer in the way he mixed 'high-level historical knowledge with social science analysis' (Shils and Finch 1949) and is therefore highly relevant in this context. His view was that the sciences are separated not only in subject matter but also in interest and the questions they pose. Social science was a science whose object was to interpret 'the meaning of social action and thereby give a causal explanation of the way in which the action proceeds and the effects which it produces' (Weber 1968 [1922]: 7). Furthermore, human action is accessible through certain mental processes; they could be understood through reliving. For Weber (and for social scientists after him) it was irrelevant whether the causes of action had a physical background, since action meant something to us only through individuals' understandings, anyway. He emphasised that 'action is "social" in so far as its subjective meaning takes account of the behaviour of others and is thereby oriented in its course' (Weber 1968 [1922]: 4). His goal for social science was a 'science which attempts the interpretive understanding of social action in order thereby to arrive at a causal explanation of its course and effects' (Weber and Heydebrand 1994: 1).

Importantly, it was not Weber's focus on meaning, social action and causes by itself that made him generally uninterested in the role of the physical environment, but the way he understood the issue. Physical geography, therefore, could enter by default, so to speak, into his descriptions for example of the state, although it is unclear how far the actors saw this relationship in the way Weber did, that is, their subjective meanings might have overlooked what Weber described as a non-social fact. He categorised states in many ways, but he also related differences between states to physical variables: (a) coastal and inland states, (b) states of the plains and (c) great river states. According to Weber, physical location also has a dispositional bearing: the coastal and maritime state offers opportunity for democracy and the state of the plains for bureaucracy (Weber 1946: 209–10). Again, when Weber was talking about 'river states' the definition was unclear but, even more importantly, he did not distinguish between rivers or explain why only certain rivers could produce a river state. He also mentioned in passing that in old Egypt, bureaucratic centralisation could never have reached the degree it actually did without the natural trade route of the Nile (Weber 1946: 213), i.e., bureaucracy was explained by the physical qualities of the river. He also made a point of what would have been the very far-reaching historical relevance (if it had

196

been correct) of the social importance of climate when he wrote that the rural social structure in England and Germany was influenced by climate in a specific way, since stock-breeding was not possible in the 'German east on account of the climate' (Weber 1946: 384). Due to Weber's overall perspective on and justification of social science, analyses or, better, descriptions ascribing a structuring role to nature (deterministic in character, it must be remarked) remained anomalies in Weber's overall conceptual approach to social science, and he never did attempt to integrate nature or the physical environment into his overall theoretical and conceptual discussions. In Weber's analysis of world religions and comparisons of the histories of civilisations, natural factors played no role whatsoever.[7] His very broad and extremely influential comparative studies of religion and his interpretation of the breakthrough of capitalism in Europe were marked by the way he defined social science and its goals, and factors of the physical environment did not enter into his interpretative universe.

Durkheim, Marx and Weber aimed to distance themselves from the discredited nature–culture determinism that was an enduring feature of Western thinking about the power of nature over human mentalities.[8] Furthermore, their understanding of the dualistic relationship between societies on the one hand and nature on the other reflected and was given status by very influential theories of historical purposiveness and the victorious march of modernity. At that time, history was regarded as a process whereby mankind gradually liberated itself from the dispositional powers of nature. The new historical consciousness that emerged reflected this struggle to escape the shackles of nature. To separate nature and society or nature and history, as the founding fathers did, was therefore in line with a deep-seated cultural and even existential tension between the Modern and the Ancient. In the third place, the nomothetic aspirations of discovering and formulating laws of social development strengthened and cemented the analytical and conceptual power of the topical focus and delineations of the new disciplines.

If knowledge systems, like other systems that seek to sustain themselves, require a specific identity, the definition of the social science project served its purpose. Research traditions that conceptualise society as a set of inter-related social variables or facts, within which those structures that should be studied are defined as belonging only to the social sphere (such as semantic or normative rules and power resources) because only such structures influence choice and opportunities, will naturally tend to refer natural factors more and more to a place outwith the analytical and conceptual universe.[9]

When the founding fathers established the social sciences, they did not or could not know that the waterscape and the role of water in the earth system were the factors that distinguished it from other planets. The limited knowledge about the universe and the planet had for centuries affected the fundamental understanding of our planet's characteristics, reflected in its naming. In English it is called 'Earth'; in Afrikaans, *aarde*; in Arabic أَرْض ('land or earth', with a definite article) pronounced as 'arD'; in Hebrew *ertz*; in German *Erde*; and in the Scandinavian languages *jord*. The English '*Earth*' developed from the Old English word *eorðe*, which means 'ground, soil, dry land'. *Tellus* is a Latin word for 'land, territory, earth', as is *terra*; 'earth, ground'. Earth got its name because it was regarded as the opposite of the wet elements on this planet, that is, the water of the oceans. Empirical knowledge about differences between continental and national hydrological situations and changes in the water cycle was usually poor at the time, and knowledge about the role of water on different planets was almost non-existent. While Adam Smith wrote *The Wealth of Nations*, Karl Marx wrote *Das Kapital*, Auguste Comte and Charles Darwin formulated their theories of biological and social evolution and, in the 1890s, when Émile Durkheim sat in his office in Paris and formulated what came to be the fundamental conceptual universe of the social sciences and, at the beginning of the twentieth century, when the other father of social science, Max Weber, was editing his *Archiv für Sozialwissenschaften und Sozialpolitik*, they did not know that the uniqueness of our planet lay in its waters and that the water balance was absolutely crucial to the human body and human physiology. There was therefore no reason why the early social theorists or historians should think about world history in terms of a water perspective or about the water cycle as being a particularly interesting factor in explaining development trajectories and social diversity.

This classical legacy of the conceptualisation of and delineation of the research object served as a very useful identity marker for social sciences *vis-à-vis* the much stronger natural science disciplines. Since their modest beginnings in Durkheim's time,[10] millions of students have been taught to study, think and reflect upon society as being produced by the social, actively disregarding or minimising the influence of non-social factors such as nature or water. If all the books on sociological and political science methodology in the University Library in Cambridge are accepted as being reasonably representative of the research interests of these two disciplines, it is telling that none of the books in the library's holdings in the spring term of 2013 offered methodological advice on how to study the relationship between ecology and society or water and society.[11] The most influential tradition has held that nature and water are of very marginal or zero interest to social scientists. The dominant theories about development that became a conceptual pillar of the global discourse on development after World War II were all based on a single fundamental conceptual aspect, be it modernisation theory, dependency theory, basic need theory or theory of rights-driven development. All were radically sociological in their point of view and all disregarded the importance or relevance of nature or water systems in explaining patterns of development, different development trajectories or strategies of development. That makes social science part of the problem it now promises to solve, not only in relation to the debate about the 'green economy', or 'sustainable development', but also concerning understanding and managing the worlds of water.

Contrary to the knowledge situation of today that makes it difficult to justify an incessant neglect of water and water-society relations, at the time when the social sciences were being established no one knew or could know that the way water flows over Tellus is our planet's defining quality. Adam Smith, Auguste Comte and Charles Darwin did not know about it, and Durkheim, Marx and Weber were not mindful of this when they wrote their most influential works on how history and society should be understood and analysed. In what Eric Hobsbawm called the age of extremes,[12] researchers and scholars were preoccupied with modernity, communism, imperialism, capitalism and urbanism, to the natural exclusion of non-ideological issues such as the role of water for social development and in historical processes. When Talcott Parsons and Michel Foucault produced their most influential works, most people were ignorant of, or uninterested in, how the unique character of the waterscape had affected societies' past and how interactions with water framed current urban life. The generation that helped to turn social science into a discipline that attracted millions of students during the 1960s and immediately after was so preoccupied with social issues such as peace and war, women's liberation, social justice, imperialism, the class struggle and Paris '68 that few grasped the long-term implications for the understanding of human history or the human predicament of what was discovered and documented later in 1968 – in space!

POST-MODERNISM AND CONSTRUCTIONISM

A concern with the physical world (or a water-system approach) became even less fashionable when post-modernism and constructionism became influential among social scientists and historians in the last decades of the twentieth century. New theoretical and philosophical arguments relegated the role of nature or non-social variables in societies' development further to the background. While acknowledging some of the conceptual and theoretical problems stemming from the well-established dualism between nature and society, post-modern trends tried to overcome them by doing away with nature altogether, both as a topic and as a term.

Nature itself was now conceived of as a social construction. It had become a *social* fact; nature was nothing if it was not social; similarly in the case of water: water is nothing if it is not social. The idea was that all that we can ever perceive about the world is shadows, and scientific knowledge should therefore not be regarded as a representation of nature, 'but rather as a socially constructed interpretation with an already socially constructed natural-technical object of inquiry' (Bird 1987: 255). In line with the assumption that nothing is knowable and that all truths are equally correct, an epistemology of nature was constructed that focused on cognitive, normative and symbolic constructions of nature. Post-modernism was a way of thinking that should be understood as primarily an epistemological assertion about our knowledge of nature rather than as an ontological assertion concerning the reality of nature itself (Proctor 1998, 2001). Nature *qua* nature was thus emptied of both natural and social significance, and in this perspective Anders' picture and all that it revealed and pointed to was ignored.

In 1993, the French sociologist Bruno Latour, almost exactly 100 years after Durkheim, published a very influential book that opposed the way in which the distinction between nature and society had been drawn. Latour's idea was to get rid of the subject–object distinction altogether.[13] Nature should not be considered 'as the external background of human and social action' at all (Latour 1999: 308). He suggested instead that there was a need to naturalise society: 'Nature and Society are not two distinct poles, but one and the same production of successive states of societies-natures, of collectives,' he argued (Latour 1993: 139). He also suggested that

> Nature and Society have no more existence than West and East. They become convenient and relative reference points that moderns use to differentiate intermediaries, some of which are called 'natural' and others 'social', while still others are termed 'purely natural' and others 'purely social', and yet others are considered 'not only' natural 'but also' a little bit social.' (Latour 1993: 85)

His aim was to highlight the cross-overs through which humans and non-humans exchange properties, taking, according to himself, the argument beyond the polemical war between objects and subjects (Latour 1993: 193–4).[14] Another related approach accepted ontologically the existence of nature but argued that it had become social or, rather, nature had become a product of socio-ecological interaction and was now a socionatural phenomenon.[15] The socionatural was therefore seen as a new, dynamic, geographical configuration where, for example, all sorts of social issues and class antagonisms could be translated into spatial configurations.

The influential German sociologist Ulrich Beck arrived more or less at the same time at conclusions with somewhat similar conceptual and empirical implications, but from another point of departure: he argued that nature *per se* no longer exists, and criticised the fallacies of positivistic scientific truths, writing that nature has been shaped by humans to such an extent that in a strict sense it is not nature any more (Beck 1995: 54). The embedding of human beings in nature is something of the past, eliminated by modern societies. Nature as something peculiar and external to the social is regarded as largely irrelevant in modern societies.[16] Beck's idea is primarily an ontological assertion concerning the 'reality of nature' itself rather than an epistemological assertion concerning our knowledge of or ability to know nature.

A common trait of the above-discussed efforts to understand nature/society interactions is the assertion that nature in itself is of no social interest or is not a variable that needs to be considered in analyses of social development, because it does not exist, whether for epistemological or ontological reasons. This book argues that water exists both on epistemological and ontological grounds, but that the distinction between society and nature must be drawn differently from in the past, not the least because water transcends the conventional boundary or distinction, and is both natural in society and social in nature.

HISTORIANS, NATURE AND WATERBLINDNESS

What about history as a discipline; to what extent is there a historiography of water/society studies? The practice of this discipline is of course extremely varied, but given the question at hand it must be reasonable to use the most influential, respected and quoted historiographies as sources for the reconstruction of recent research profiles of the discipline.

Iggers and Wang's *A Global History of Modern Historiography* (2008) is widely regarded as an authoritative summary of the past century of global historical research. Significantly, in this book that discusses how historians have conceived of their discipline and explained historical developments, there is not one word about nature or environment or ecology. Reconstructing the practice of modern historians, it contains not a single reference to scholars who have grappled with how to analyse the relationship between nature and society or who have offered empirical descriptions of that same relationship. Of course, historians have produced empirical studies about the Hoover Dam, British canals and Norwegian hydropower, and about the Ice Age and adaptation, but no regional or national histories have been written putting water-society relations in focus and when the historiography of the discipline is written, studies of relations and interactions between nature and society are not included.

Other influential works on historiography confirm the clear impression and picture given by Iggers and Wang; Daniel Woolf's *A Global History of History* (2011), *Writing History: Theory and Practice*, edited by Stefan Berger, H. Feldner and K. Passmore (2003), *Making History* (2004) by P. Lambert and P. Schofield, and Kenneth R. Stunkel's *Fifty Key Works of History and Historiography* (2011) all have no entry on nature, ecology, environment, water, rivers or physical relationships, while the last book does have two entries on nature and two on laws of nature but does not deal with nature as a factor in establishing and developing patterns and histories of interactions.

The evidence makes it quite easy to conclude: during the last century – when the relationships between nature and society and between water and society were fundamentally transformed and such issues as resource use, climate developments and water control were at the heart of development and social engineering – historians scarcely studied nature/society interactions and at best integrated them into the analysis as a static backdrop to human action. There are many reasons for this situation. It is a logical consequence of the methodological stress on written texts as the proper source material of the historian to the detriment of interest in technological artefacts, rainfall statistics, geo-morphological changes in the landscape and so on. It is of course also partly a reflection of dominant thinking about nature and water in modern, urbanised societies, since the Age of Modernity has also been an age of general waterblindness. It is also just one of many signs of how history too has been influenced by fundamental justifications of social science in its early days, and by the ideas of Marx, Weber, Durkheim and others.

But history has its own theoreticians who have pulled in the same direction. Here the Oxford historian and philosopher Robin George Collingwood (1889–1943) will be discussed. He has until the present day exerted a very strong influence via his classic works *The Idea of History*, published after his death in 1946 and following another posthumously published book, *The Idea of Nature* (1945), and his *Principles of History* (1999), and the way he argued has resonated widely in historical departments all over the world.

Collingwood argued in favour of the same distinction between 'Naturwissenschaften' and 'Geisteswissenschaften' that dominated social science after Durkheim. He arrived at this standpoint by reflecting on what was meant by 'history', being influenced along the way by Kant, Hegel and Vico. He emphasised that thinking about historical thought is essentially thinking about the object of historical thought. 'All history is the history of thought' Collingwood 1946: 215, 304). However, history deals with thought in general and with 'actions done by reasonable agents in pursuit of ends determined by their reason'. And for a man about to act, 'the situation is his master, his oracle, his god' (Collingwood 1946: 316). All history is the history of thought, and the crucial thoughts are those that deal with understanding of 'the situation'. The question that the genuine historian asks is not what kind of event usually precedes the event that is about to be explained, but what reasons make the action intelligible? According to Collingwood, this 'situation' should be analysed without asking questions about how different natural conditions framed situations in different ways. Since the historian should deal with or be concerned with rational connections rather than inductive generalisations, nature becomes both theoretically and empirically irrelevant, even as a background factor.

My point is that we do not need to resort to inductive generalisations or writing 'pseudo-history', as Collingwood calls it, by bringing natural factors into the picture. Collingwood overlooks or is forced to neglect the fact that the natural environment and modified landscapes create and reproduce different 'situations' for those involved, and thus make only some actions rational and some thoughts possible and intelligible. For example: for entrepreneurs to plan to establish factories based on waterwheel technology was a rational idea in parts of England but would have been madness along the Nile in Egypt or in the Sahel belt in Africa. The same action and the same intentions within similar systems of thought would have very different implications. To worship water as a gift of God was a more intelligible ritual in the deserts of Arabia and Palestine than on the Scandinavian raincoast. In *The Idea of Nature*, Collingwood discusses not the history of ideas about nature in general, although he describes the book as though he does so, but the ideas of nature in the specific context of Western Europe's climate and waterscape. His failure to grasp how ideas about nature and individual aspects of nature also reflect the existence of different natures also demonstrates the problem with his ideas about history.

Collingwood is one of the few historians who have been interested in and written thoroughly about nature, also in a theoretical perspective. He did so in congruence with his general programme for what history is: a history of ideas or of human thinking about nature and not a study of how nature as a factor or variable impacting on the situations his actors are acting on or within. It is also highly relevant from our perspective that he conceives of nature as one entity. He relates the development and change in thinking from the Greek cosmology, which regarded nature as an organism with intelligence, to 'Renaissance cosmology', where nature is operating according to external laws stemming from the divine creator or ruler of nature, and is based on a clear distinction between mind and body. He argues that, in the nineteenth century, a more historical understanding of nature gained ground, especially in the form of different theories of evolution. Collingwood regards nature as a whole, and the history of the ideas of nature that he constructs is seen solely as a product of other ideas and is never analysed as being affected by the nature of the different natures these ideas develop within. The reason why it becomes possible for Collingwood to write a history about the ideas of nature without bringing different natures into the picture is actually twofold: he restricts his analysis to Europe, throughout which natural conditions are quite similar in many fundamental aspects and can thus with some justification be treated as a constant, omnipresent factor that becomes, in the course of time, irrelevant as a factor in explaining different trajectories of development. His approach to nature as a unitary phenomenon leads naturally to a general conceptualisation of nature as an idea rather than as an experience, since in the real social world actors and people experience aspects of nature and never nature as a totality or unity.

In social science it has been conventional to argue that there are four logical possibilities for conceiving the relationship of nature and society: the reduction of society to nature, the projection of nature into society, dualism, and a nature-society-dialectic. This book has argued for a fifth and more fruitful possibility for conceiving the relationship between nature and society, at the same time overcoming rigidities and dichotomies inherent in these approaches.

NATURE DETERMINISM AND WATER

The second section of this part of the book will locate the water-system approach in relation to a critique of some of the social scientists and historians who distanced themselves from the dominant tradition and who sought to integrate non-social variables such as nature and water in their analyses of societies. This is done because the most famous contributions of this counter-hegemony trend have been strongly criticised for their nature determinism, and it is therefore thought useful to compare the water-system approach with these contributions. The approach suggested here is not deterministic but provides rather a means to overcome the fallacy of both reductionism and nature determinism, and far from being suggested as a substitute for other explanations it is pushed forward as an overlooked but crucial addition to existing narratives.

The geographer Ellsworth Huntington aimed famously to explain the development of human civilisations as a result of their interactions with nature and climate. Huntington has been criticised for determinism and for having a superficial understanding of the complexities of societies. I concur with the criticism of his way of discussing race; the relationship between climate and mentality; and how he used certain aspects of nature to explain social differences. However, this well-known and relevant criticism will not be repeated here. Instead, his work is analysed from a very different angle. The point that is made here is that his understanding of nature and his methodology were more fundamental to his determinism than previous criticism has realised.

Huntington was convinced of the unity of nature and of the 'unity of science'. Subsequently he did not comparatively study particular aspects of nature in their empirical interactions with societies in detail. His model of the role of nature in its relationship with society was therefore basically static. His view on nature and society therefore forced him to study 'everything'. Huntington wrote in line with this that the 'unity of nature is so great that when a subject such as climatic changes is considered, it is almost impossible to avoid other subjects, such as the movements of the earth's crust' (Huntington and Visher 1922: xi). His book therefore discusses everything from the causes of earthquakes to how climatic changes may be related to great geological revolutions in the form, location and altitude of the land and all kinds of potential physical factors which have moulded the evolution of organic life, including man (Huntington and Visher 1922). In his most famous book, *Mainsprings of Civilization* (1945), he aimed to 'analyze the role of biological inheritance and physical environment in influencing the course of history', this being part of a yet larger plan which included 'an interpretation of the main trends of history in the light of these two factors as well as of the cultural factor, which is generally the main topic in histories of civilization' (Huntington 1945: 1). These quotations can serve to illustrate the enormous scope of this project – all of nature, all societies of the world, all of the time. It was rationalised by his unitary view on science and nature. Huntington could not manage, of course, in any comparative, empirical way to study all the actual interconnections between societies and nature. It was therefore to be expected that he should suggest some rather sweeping and basically deterministic explanations of social developments.

Huntington's view of nature brought natural factors into the equation, but the way he approached and understood nature made it difficult to study actual interactions between aspects of nature and societies in a rigorous, comparative way. He mentions both water and rivers (never water systems and river basins) but does not describe water and rivers systematially or discuss how individual rivers or waters influenced particular development patterns and challenges.

Typical of his approach is the formulation that such 'details' as the differences between braided rivers and meandering rivers are of no interest to people concerned with human development (Huntington and Cushing 1922: iii). In reality, however, such apparently minor divergences proved to be of tremendous importance, for example in the diverse patterns of development of the industrial revolution, as they can partly explain the different situations in which entrepreneurs in various countries and regions found themselves in employing new technology at the end of the eighteenth century (see Chapter 2 of this book). Of course, meandering rivers did not create the foundations of the sorts of regular power sources and trading routes that modern capitalism and production required to thrive. Huntington also jumped to rather general conclusions about water/society relationships, based on what must have been limited and biased empirical data, describing, for example, rivers one-sidedly as only 'barriers' to trade and development (Huntington and Cushing 1922: 130–1), rather than also trade routes or highways to the world market, as he did in his later works (Huntington 1945). The point is that major river systems, due to their different hydrological characteristics, played various roles: in China they were both a highway and a barrier, while in present-day northern Pakistan the Indus really was the most important barrier to trade and exchange, and in England, rivers like the Severn and Thames were among the most important trading routes in the country.

Huntington today is primarily remembered for his natural determinism, which in some cases assumed a racist overtone.[17] Criticism of his attempts to explain sexual matters among the Zulu by tropical heat, for example, has naturally tended to overshadow a discussion about more theoretically important aspects of his conceptual and analytical approach.[18] He failed to be precise in identifying which aspects of climate he studied, for what reasons and in which societies. His assessments of impact overlooked both the complexities of the interactions of specific societies with the physical world and the importance of the latter to them. Instead of studying aspects of climate change and how they impact on particular sectors or activities in society in different ways, he conceives of climate as something reduced to a form of general 'social power'. It is as this form of power, and not as nature *per se*, that climate becomes the mover of this determinist causation.[19] By focusing on water as one aspect of the climate, it will be possible to study precisely how this relation between climate and sociey is multidirectional and much more open-ended and dependent on social response and pro-activity.

Based on the water-society system approach we may rather say that 'climate' has influenced particular environmental adaptations without having determined historical development; it has influenced where human beings can thrive and live, what crops may be grown, and where cities should be located, but has not decided where cities are, what crops are grown and where human beings thrive.

Other attempts at integrating nature or ecology in social analyses have been suggested by influential athropologists. One reason for this discipline's interest in the role of ecology is that anthropologists tend to study relatively underdeveloped agrarian or hunter-gatherer societies. They have researched relationships between society and environment in cases where the link is evidently important and apparently easier to study.

Julian Steward and his 'cultural ecology' approach is the most influential proponent of this approach. In his most influential contribution, he traced evolutionary similarities in five ancient civilisations: Mesopotamia, Egypt, China, Mesoamerica and the Andes. According to Steward, these cultures shared parallels in their development of form and function. The main reason was that all of them developed in arid and semi-arid environments whose economies were based on irrigation and floodwater agriculture. He argued that these social similarities stem not from universal stages of cultural development, nor from the diffusion of civilisation between these regions, but from their similar natural environments.

Steward aimed at demonstrating that evolution occurs along parallel lines (multilinear evolution) that are determined by differential environmental adaptation (Willey and Sabloff 1980).[20] He was looking for a methodology that could determine regularities of form and processes which recur among societies in different cultural zones (Steward 1972: Vol. 1, 3). He was not looking for a universal theory of evolution but for a theory of evolution of parallels of limited occurrence. Steward argued that different cultures share evolutionary features that can and should be explained as parallel adaptations to similar natural environments. On this basis, he formulated a theory of social evolution that explained social systems in terms of their adaptation to environmental and technological conditions. Or as Steward wrote: 'The cross-cultural regularities which arise from similar adaptive processes in similar environments are functional or synchronic in nature' (Steward 1972: Vol. 1, 5).

The main problems with this 'cultural ecology' approach are twofold: (a) it is not very helpful in analysing modern, more complex societies in which mechanisms of adaptation are much more multidirectional and in which interactions between societies and nature are much more complex, and (b) it is deterministic in the sense that it is not really concerned with the differences in the irrigation system in the areas he compared, and especially in the water systems that provide the foundations of various irrigation systems. The lakes and artificial gardens of the Aztecs, the seasonal rain-fed dams of the Mayas, and the annual floodwater of the Nile in a country where it almost never rained filling the basins along the river are extremely different ecosystems. The agricultural practices they encouraged were also very different. The similarities in social structures that Steward asserts are there can thus not be explained by natural factors in the way he does. It is therefore an assumed connection, and since he explains these similarities with nature, it is a form of anticipated, deducted determinism. The environment and waterscapes of Egypt, the Andes and China in particular are much more different than the theory of limited parallel development acknowledges. It is precisely by highlighting the great variety and constant flux of different water cycles and waterscapes and the quite advanced knowledge we have of the different agricultural civilisations and their relations to their waters that it will be clear that such general, partly deterministic models of social behaviour must be falsified.

The political scientist Karl Wittfogel in his book *Oriental Despotism: A Comparative Study in Total Power*, forcefully reopened the question of the relationships between modes of production and environmental conditions. He attempted to understand and interpret the abundance of dictatorial regimes in basically geographical terms and saw these political and institutional conditions as direct or indirect consequences of irrigation agriculture in large river basins in dry areas. He generalised his notions of 'oriental despotism' to every dynastic empire with a river running through it – China, Russia, Persia, Mesopotamia, Egypt, the Incas, even the Hopi Indians of Arizona. The origin of despotic government was to be found in the initiation, implementation and operation of large-scale hydrological works.

The typical hydraulic empire government, according to Wittfogel's thesis, is extremely centralised, with no trace of an independent aristocracy. Hydraulic hierarchies gave rise to the establishment of strong, rather permanent institutions of impersonal government. Wittfogel's model has been harshly criticised. Joseph Needham argued essentially that Wittfogel was operating in ignorance of basic Chinese history, arguing that Chinese governments were not in general despotic or dominated by a priesthood and experienced many peasant rebellions, proving that dissent was possible. Wittfogel's analysis of Egypt has likewise met strong opposition from archaeologists and other scholars, who have shown that irrigation started without being dependent on a strong state and a weak society.

I concur with the criticism by Needham and others of Wittfogel's work, but here a very different point is made that is thought to be methodologically more fundamental. The main problem with Wittfogel's thesis is essentially that he was not really interested in mapping the characteristics of the rivers he claimed were 'responsible' for the social systems he describes. His book contains no detailed descriptions of their hydrology and the way they ran through the landscape. In fact, Wittfogel wrote as though the Yellow River and the Nile and the Euphrates-Tigris were similar rivers, or at least similar enough to produce similar types of institutions and societies. Wittfogel mentions the 'specific qualities of water': its susceptibility to movement and the techniques required to handle it, that it is the natural variable *per se* in a given agricultural landscape, and that it flows automatically and according to gravity (Wittfogel 1957: 15). But more importantly, he does not describe in sufficient detail the different river systems and how they change over time and pose specific managerial and political problems. He implies that the river systems of Egypt, Mexico, Turkestan and China were similar, without specifying to what extent they really shared important characteristics (Wittfogel 1957: 24). He also suggests that only irrigation civilisations were interested in canal building, stating that canal building started in Europe with the Canal du Midi and overlooking much earlier water modification schemes for both agricultural and non-agricultural activities in the Netherlands and in England, and in connection to the Hanseatic cities long before the French decided to link the two oceans by canal. The weakness of Wittfogel's thesis was therefore not that he was concerned with how rivers produced social relations, but that he was not sufficiently interested in how different types of rivers help over time to develop diversity in social responses to river systems.

A last example can be the physiologist Jared Diamond. In *Guns, Germs, and Steel: The Fates of Human Societies* from 1997, he asks the very appropriate question: 'Why did history unfold differently on different continents?' His answer is: different environments, rather than culture or mentality. Diamond puts a great deal of weight on diseases and advances the hypothesis that successful cultures are those that settle continents aligned along an east–west axis. The only cultures that fit this description are Europe and Asia. He makes environmental conditions a cause of historic development.

His model fails to explain the overarching question: why did Asians not colonise the world or organise an industrial revolution in the eighteenth and early nineteenth centuries? He mentions several factors, and although he ascribes the rains of Africa a very important role in the history of that continent he does not carry out more systematic and rigorous comparative studies of the role and character of rivers, the seasonality of precipitation, scale of evaporation and so on and how particular water environments over time helped to establish a foundation for a practical-scientific milieu of entrepreneurs in parts of Europe. This omission is clearly reflected in the way he has organised the index, since it has only one entry directly related to water and that is water power. According to the book and the index water power is on a par with watermelons and water buffalo, the other entries related to water. He has no entry on rivers, none on mills, one on irrigation (irrigation systems) and one on hydraulic management. The last two entries deal with the same point: Diamond criticises the 'hydraulic theory' of state formation because it addresses only the final stage in the evolution of complex societies and this theory says nothing, he argues, about what drove the 'progression from bands to tribes to chiefdoms during all the millennia before the prospect of large-scale irrigation loomed on the horizon' (Diamond 1997: 283). Instead of what is described as a deterministic version of the 'hydraulic theory' he suggests his thesis of population pressure.

Diamond's criticism of the hydraulic thesis is fruitful, but his comparisons of the Mayas, Aztecs, Chinese, Madagascans and Egyptians are too general. It is necessary to analyse systematically the different waterscapes and their evolution through history and how they have encouraged different development trajectories, or how different societies have adapted to and controlled their waters. Such an analysis would have been less deterministic and easier to check. Diamond objects, however, to his analysis being deterministic: 'It's just that some environments provide more starting materials, and more favorable conditions for utilizing inventions, than do other continents.' (Diamond 1997: 408) The point about the water environment as compared to Diamond's environment is that this is not only a given, quite fixed, external structure, in the way that Diamond conceives of environmental conditions. The waters, or the rivers and the streams and the run-off, can on the contrary be controlled, changed and exploited in very different ways at different junctures in history. A difference and a problem is also Diamond's continental scale as conceptional unit, since variations in the behaviour of the water cycle or in river systems on a specific continent will often be greater than differences between the continents. The consequence is that similarities in climate or water systems must be assumed or ascribed, as is implicitly the case with the theory of the north–south axis.

In his next book, *Collapse* (2005), Diamond suggests that climatic and environmental determinants have been the central factor in the rise and fall of empires. This book too has been criticised for environmental determinism because he argues that in societies where most control was exhibited this was frequently due to the central role played by limited resources in economic processes. This constrained nature made control of supply and demand easier and allowed a more complete monopoly to be established, as well as preventing the compensatory use of alternative resources. The analysis would have been less deterministic had it paid more attention to concrete changes in local and regional climates and water cycles and run-off, since to a large extent Diamond explained social changes in terms of natural phenomena such as droughts in the Mayan region and changes in the form of water on Greenland. A comparative and thorough empirical analysis of the collapse of the Sumerian and the Indus civilisation, and the fate of Pharaonic Egypt and classical China, will show that the relationships between changes in water cycles and social development cannot be reduced to a simple one-directional causal one.

WATER AND THE 'TRAGEDY OF THE COMMONS' – A CRITIQUE

Since this book is arguing that the water issue and the importance of the water-society nexus have basically been ignored in mainstream social science and history, it opens up an extremely wide array of empirical studies – both of new topics and questions and of old topics and questions in new contexts. The dominant waterblindness has, moreover, also exerted a major influence on theories of social development and social action, not only because the water factor has been overlooked but because water has characteristics that will tend to erode theories and concepts formulated as if water is not found on the planet.

The first question that can be reappraised is a famous one: how can we optimally share common resources? This general question has given impetus to an extensive literature that has had a profound impact on social theory and thinking about society in general. A basic assumption is that common property resources have long been overexploited and misused by individual players acting in their own interests. A long line of collective action theorists has argued that people placed in a situation in which they could all potentially benefit from co-operation are unlikely to co-operate in the absence of an external enforcer of agreements. An equally long line of property rights theorists has suggested that common property resources are bound to be over-exploited as demand rises. Of course this has also been a major challenge regarding the allocation of water resources among national and regional stakeholders, which has produced a number of influential analytical models for understanding river basins all over the world.[21] Our discussion of the concept here aims to demonstrate how waterscapes' particular form of territoriality and the concept of river basins that also creates a nexus of social relationships questions the validity of the most influential model, that of the 'Tragedy of the Commons'.

The most famous model invoked to explain behaviour and solve problems related to the management of shared resources is that proposed by Garrett Hardin (1968), often referred to as the theory of the 'Tragedy of the Commons'. Hardin located his actors on a pasture shared by herders, where each individual herder acts rationally and wishes to maximise his yield. But each additional animal introduced to the pasture has both positive and negative effects. The herder increases his return but, meanwhile, the pasture is degraded. By 'the remorseless working of things', the rational actions of self-interested individuals do not promote the public good, and in the end they will also negatively affect the herder who first increased his herd. Hardin wrote: 'Therein is the tragedy. Each man is locked into a system that [causes] him to increase his herd without a limit – in a world that is limited. Freedom in a commons brings ruin to all'.[22] The tragedy of the commons arises because resources will be exhausted by rational, utility-maximising individuals rather than conserved for the benefit of all. And the greediest herdsmen would gain, but – and that is the point – only for a while, the reason being that mutual ruin is always around the corner. The unmanaged commons will be ruined by overgrazing caused by interest-maximising individuals. Pessimism about the possibility of users voluntarily co-operating to prevent overuse has been widely used to justify both the privatisation and nationalisation of commonly owned resources. The model pasture is ecologically quite uniform, as are the economic activities of the herders and the available rational adaptation mechanisms and strategic choices of the actors who share it. The pasture is the model's ecological environment, but if adaptation mechanisms and strategic choices are different in other ecological contexts, Hardin's model cannot be as universally applicable as once thought.

Two main objections can be raised against this theory based on the water-system approach and an understanding of the characteristics of large rivers and water's confluences with society.

A pasture, as a physical space in nature and thus also a resource, is fundamentally different from a river. Actors operating in other physical resource contexts, such as an international river basin or a large national river basin, might therefore think about cost and benefits in other ways than the pasture-based model predicts. Actions motivated by similar intentions and norms might also have very different implications in different physical settings.

There are waterscapes that have traits that are comparable to those of a pasture, and with regard to groundwater basins, Hardin's theory is instructive. Nobody really owns the groundwater, and it is technically 'up for grabs'. History is full of examples where individual and rational pumping of groundwater has resulted in the depletion of the resource for all. That has led to other losses, most often as saltwater intrusion, very revealingly on parts of the Indian coast and in the Nile Delta, and as land subsidence, as in Mexico City, in Venice in 1984, when the authorities stopped the city from sinking by prohibiting the pumping of groundwater from beneath the seabed. In the ecological context of an aquifer, diffusion acts to spread the effect of the individual's use among all. The mechanisms involved are very clear: depletion by a few means depletion for all. The example shows that it is not water as a physical element in nature *per se* or as a resource in itself that objects to the theory of the tragedy of the commons. It is rather the process of running water over large distances that may cross several ecological zones and that coincide with and help to develop different water needs and water usage practices that does that.

Most rivers, particularly long rivers, are ecologically extremely varied. For example, the Nile traverses three major climatic zones from tropical Africa with its rainforests, crossing the savannah and running through the Sahara without receiving a tributary for 2,000 kilometres before emptying into the Mediterranean. The Nile basin covers one-tenth of the continent and is part of 11 countries with a population approaching half a billion. Such rivers lend themselves to a wide range of strategic choices and economic adaptations at various points along their courses and among actors occupying different water-related ecological niches.[23] Geography and scale might also be factors explaining differences in water strategies and ideas about water. The theoretical configuration associated with pasture ecology will therefore not be reproduced in the context of a river basin, owing to its physiography, topography and the resulting unequal structural positions of actors in relation to the resource.[24]

Unlike for pasture, the rather diffuse and shifting boundaries that are a distinguishing physical characteristic of river basins or watersheds have immediate implications for co-operative frameworks and thus for the whole tragedy of the commons model. For example, if countries located upstream in a large river basin develop rain-based agriculture, when and to what extent should use of this water that would otherwise drain into the river be considered part of the common resource or the shared watershed? A more controversial question is whether or to what extent a river basin should be defined topographically rather than according to where the waters actually run. Can, for example, the Egyptians legally and as part of their co-operation with upstream countries pump water across the Suez Canal and into Sinai and can they pump water up from the High Aswan Dam into the Sahara Desert in the Toshka area to create an artificial Nile valley there, and still claim that they are using water within the boundaries of the Nile basin? Egypt argues that these areas belong to the Nile basin historically, although water no longer reaches them, mostly due to human control and modifications of the river. Egypt objected on its side to Tanzania pumping water from Lake Victoria to cities in the Shinianga area to the south of the lake, areas which Tanzania claims is draining water back to the lake. And, in Ethiopia, water planners have long discussed schemes for sending water in tunnels to dry areas outside the Nile basin proper. Because of the physical character of water, and the fact that the actual boundaries of a river basin are always in flux, the whole issue of 'clearly defined boundaries' creates a divisive element precisely because they are not clearly defined.

One riparian actor, be it a state, private companies or individual farmers, can in large river basins use or even control the river to maximise yield without negatively affecting other users or the river itself. Egypt, for example, exploited the River Nile for thousands of years without any harmful or other effect on present-day Kenya or Tanzania. Even during the past few decades, the country's extensive exploitation of the river has been of negligible importance for its upstream neighbours, since their lack of development of the river was primarily the result of internal development, lack of political stability and technological capabilities rather than of Egypt's use of the resource.[25] In river basins, moreover, pursuit of self-interest upstream can *benefit* downstream users. A hydroelectric dam, for example, may protect downriver areas against floods and reduce the problem of silting. The Roseiris Dam in the Sudan would have had very negative effects had it been built during those millennia when seasonal flood irrigation dominated Egyptian agriculture, but after the High Aswan Dam was opened, the benefits of Roseiris far outweighed the disadvantages for Egypt, because it trapped Blue Nile silt, thus protecting the Aswan reservoir. The natural domains and the right regimes of resources alter with time, technological progress and the circumstances of stakeholders. Rivers thus fulfil different demands at different places and times, and it is these natural characters that can encourage co-operative action seen as rational individual behaviour. The relationships are more subtle than Hardin's model postulated, because of river hydrology and physical characteristics and the ways in which these produce diversity in man–river relationships.

Rivers, it should furthermore be recalled, unlike the pasture in Hardin's model, always change (in some cases also dramatically) and will continue to do so, not only as a result of human interference but also as a result of changing rainfall patterns, evapotranspiration, cloud formation, atmospheric pressure and so forth. Permanent insecurity and endless fluctuations from year to year and from season to season guarantee that an individual (or state), acting rationally to maximise yield in certain cases might opt for voluntary co-operation to achieve joint control of the river because it is impossible to manage this alone. The point is that the physical world thus presents social science with a set of variables that should not be overlooked in analyses of social actions in specific resource environments.

The way the tragedy of the commons concept fails to take power relationships into account is also a crucial problem of the model as such since it attempts to explain also management practices in large watercourses that cross national boundaries and climatic zones. In such often enormous physical spaces, people have developed a wide variety of resource adaptations, and in some cases these different water-society relations and systems have even helped to form different types of states and political and economic systems. Just one example will briefly be mentioned here: Egypt, with a strong and one of the most stable state institutions in the history of the world, shares the river with Southern Sudan, most likely the weakest state in Africa and where anthropologists have always gone to study the most extreme stateless societies.[26] The state actors, or the equivalent to the herders in Hardin's model, share the same resource – the river – but they tend to conceive of themselves as living in different worlds and also in different 'water worlds'. Rivers, with their diverse hydrological and physical attributes, encourage different ways of relating to the shared water, and stimulate a wide range of economic activities connected to, and ideas about, the common resource. The classic dilemma of a dominant individual incentive that creates a suboptimal social equilibrium is therefore less likely to emerge in such river basins.

Differences in technological capabilities and other social factors as well as physical location along the river affect the ability to participate in collective action. In any river basin, the distribution of both costs and benefits is decidedly skewed, and will continue to be so as long as people live where they live, because their notions of place and space are influenced by their location within the river's overall physiography, hydrology, topography and longitudinal profile. Transboundary watercourses do not constitute common pool resources that can be exploited jointly and simultaneously and this also makes the tragedy of the commons proposition unsuitable; the pasture and the basin are different types of entity.

A related issue is the puzzle of water and property definitions. The problem of defining and delineating property in water raises issues of an even more general nature. Although water is the most important resource of all it is often overlooked or neglected in general discussions on property. In a recent influential book the water issue is typically ignored. *The Law on Property* by Lawson and Rudden states first that not everything is property: 'For something to be property it must in the eyes of the law be capable of being appropriated, so the air, the clouds, the high seas, are not, for legal purposes, property' (Lawson and Rudden 2002: 220). They continue: 'The most obvious distinction among tangible objects is between those which are (more or less) immovable and those which are (more or less) movable [...] The class of immovable is called real property; everything else is called personal property or personality' (Lawson and Rudden 2002: 22–3). Tangible objects include land, living creatures and goods, while intangible assets are commercial paper, stocks, shares and bonds, etc. Neither international rivers nor running water are covered in their discussion, nor is water even mentioned. If it had been, it would have undermined their basic distinction between immovable and movable goods, because water is both movable and immovable.

This duality of water bodies has affected efforts to define flowing water in legal terms. First, there is the problem of assigning territorial rights to objects that are both stationary and non-stationary. The argument that a river must be treated as a functional whole and in this sense is stationary is based on two false assumptions. First, what is described as the intimate causal connections between numerous uses along the river, down-, cross- and upstream. This is a normatively relevant argument and very important in international law, but the extent to which it is a reality is a question that needs to be assessed concretely, since it is far from being a universal condition in river basins. What Sweden has done with the Klara River in Sweden has had no effect on the same river in Norway and what the Dutch did with the Rhine in the delta had no implications for the river's upstream neighbours. The extent to which upstream use of the river affects downstream users is an empirical question that requires analyses of the kinds of interventions involved, of local hydrological and topographical characteristics and so on. Secondly, there is the idea that a river is a kind of 'immovable' object because its banks (although not its waters) are stationary. This assumption cannot be accepted since very few rivers follow exactly the same course year after year. Most rivers change direction and velocity, at least if left to themselves, due to meandering and the physics of moving water, but also due to human intervention, changes in climate, erosion, and so on. Rivers are drawn on maps, but the path of a river can change dramatically over time and maps constantly become outdated. India is a case in point on a grand scale. The sub-continent is called the sub-continent of 'forgotten cities' because the rivers change courses and leave behind cities established on their banks (Wood 1924: 3). The Huang He has regularly shifted its course, sometimes, in the nineteenth century, by more than 1,000 kilometres from one year to the next! The Semliki River, a Nile tributary, has been the borderline between Uganda and the Democratic Republic of Congo. When it started to change its course it moved the national borders, since the border lies in the middle of the river. It is expected that in the very long run, the Brahmaputra may dramatically change its course and perhaps no longer flow into India. On a smaller scale, streams change course all the time and because of these natural characteristics produce all sorts of social conflicts and co-operation challenges between neighbours and farmers. Rivers are far from stationary even if they are defined by their banks, and are therefore not stationary topographical phenomena. The fact is that it is not only their constituent elements that are on the move, but often also the rivers themselves.

A river cannot be defined as a stationary object, since the water and also its banks are not parts of an organised whole that maintains and reproduces the river's functions, because the functional organisation does not necessarily remain the same.[27] Rivers often continue to water their surrounding environment in the same way year after year, but in very many and important cases rivers do not follow such paths of regularity. In fact their geo-stationary features and functions vary, and since rivers are far from being topographically constant, claims of river control cannot be based on the idea that they are geo-stationary objects. Rivers have banks between which water flows, and in most cases they run from inland to a sea, but not always, and their velocity and water discharge will of course also differ in time both due to natural and social causes. Their waters have an average, approximate depth and speed, but these values change dramatically from place to place and from year to year. But, although they are not geo-stationary objects or 'immovable objects', their waters are amenable to physical division. It is also possible to divide a river, and separate essential elements of the functional organism from the organism itself. Intra-basin water transfers have taken place for centuries in many places, and also on a grand scale as in Sri Lanka, where for hundreds of years rivers have been artificially linked to other rivers in other watersheds, making it necessary to think more deeply about the different physical and man-made levels that constitute a river basin or a watershed. A river is not like a living organism, and therefore it is a divisible mass. The interconnected functions of disparate parts of the river do not resist artificial division.

The point is that, contrary to common wisdom, down- and cross-stream states can exercise vital unilateral jurisdictional authority over their riparian regions without co-operation from up- and cross-stream states. States can also and do exercise unilateral jurisdictional authority over their own section of the river. One can argue on ethical grounds that states should not possess this power, but one cannot argue that riparian states are unable to exercise or incapable of exercising their rights without the co-operation of other states, because of the nature of rivers. There are many good reasons for shared jurisdictional authority, but this 'natural' argument does not 'hold water'.

The characterisation of flowing water as common property generally places it outside the range of things that are capable of being owned. While this is true from one point of view, it is not the characterisation that is the point, nor the social construction of water as a resource. It is the resource itself and its characteristics in nature and in society that evade the usual definitions of property, and it is precisely this that makes this especially interesting and challenging when management issues are involved.

THE ENIGMA OF WATER AND THE QUESTION OF TEMPORALITIES

The water-system approach may also be useful for discussions on how to handle the temporal dimension empirically in social science and history. The nomothetic idea that science should formulate universal laws on the one hand and the idiographic conception of history as a record of particular events on the other both tend to ignore the problem of the temporal dimension (Wallerstein 1998). Given the aim of establishing social laws, the nomothetic tradition is concerned with neither the long nor the short term, since time is basically either a problem, irrelevant or an analytical residue. Historical research primarily concerned with particular events does not naturally require reflections on temporal dimensions. In dealing with particular topics or events located within the 'eternal' relationship between water and society and the diversifying influence of waterscapes on social structures over long periods of time, both approaches are clearly inadequate and reflections on time become a must.

Notions of time and the unequal chronologies of societal and natural developments have laid the foundations for the belief that societies and social change are influenced by social variables alone. Human beings have existed for about half a million years and Homo sapiens for about 100,000 years and during all this time they have been forced to relate to the waterscape surrounding them. For most of this period very little happened in terms of social development or control of water, but the water jar was invented which was important for the development of settled agriculture. Agriculture which required some form of active interaction with the waterscapes is only some 10,000 years old, and the classic agricultural civilisations based on irrigation, natural and artificial, date back no more than about 6,000 years. If we think of the entire span of human existence as a 24-hour day, settled agricultural relationship and its new relationship to water would have come into existence at 11.56 pm, the irrigation civilisations in Asia and the Middle East at 11.57, while industrialisation began to develop only at 11.59 and 30 seconds. Yet perhaps as much change has taken place in the last 30 seconds – when the hydrological cycle's geographical variations have changed very little – as in all the time leading up to it. Compared to natural changes in the hydrological cycle in the same period, the links between social development and the physical environment seem therefore to be very weak. However, the eternal laws of nature and of the hydrological cycle have laid the foundations for the diversification of social development, and they manifest themselves within human lifespans, and even minor changes in the way water runs through nature and societies often had major social, economic and political consequences.

One way of discussing the fruitfulness of the water-society system concept in this context is by approaching it via Fernand Braudel's ideas about time. One of the most influential attempts to address the temporal problem in history and the social sciences is that of the French historian. Braudel argued for the usefulness of thinking in terms of different timescales. He wrote:

> On the surface, the history of events works itself out in the short term: it is a sort of micro history. Halfway down, a history of conjunctures follows a broader, slower rhythm [...] And over and above the 'recitatif' of the conjuncture, structural history, or the history of the longue durée, inquires into whole centuries at a time. (Braudel 1980: 74)

Braudel's famous suggestion was that we should deal with three timespans of history: events, cyclical movements and *longue durée*. He also discusses the '*extrême longue durée*', a potential fourth timespan.[28] The second foundation for his notion of time is that of 'simultaneity', which combines the past, present and future in the historical examination of objects (Santamaria and Bailey 1984: 79). The effect of this way of thinking about the temporal dimension in historical and social studies was to enable analysts to 'dissect historical time into geographical time, social time and individual time' (Braudel 1980: 21).

Braudel has been criticised for not being able to reconcile his three time frames or establish clear relationships among them, as well as for a kind of determinism, being primarily concerned with factors which inhibit transformations or the notion of the exteriority of social facts which constrain and channel human action. His ideas of time will be better understood if they are seen as having been developed in connection with his study object and the particular temporal relationships between the different elements in his narratives. He worked out his notions of temporalities in his masterpiece about the history of the Mediterranean world during the age of Philip II, as is also reflected in the way he divided his story into three main parts: the role of the environment, collective destinies and general trends and events, and politics and people. This division of 'different times' was perfectly logical, given Braudel's focus on one of the world's oceans as the ecological and geographical determinant. The Mediterranean is an object of the physical water world, but is characterised by its relative changelessness. It did not change at all in the historical period analysed by Braudel.

In Braudel's narrative, geography is therefore by definition synonymous with the long term, and it had a kind of foundational implication for his understanding of the relationship between environment and time in general. In contrast to this geographical time Braudel discusses the

219

'resounding events' in societies, described as 'momentary outbursts, surface manifestations of these larger movements and explicable only in terms of them' (i.e., in terms of longer-term social factors or structures) (Braudel 1980: 21). If instead we focus on freshwater systems, rivers and rainfall in the local hydrological cycle, that is, on elements in nature that are characterised by being both eternal and always in flux, and also strongly influenced, in both the short and the long term, by both humans and natural forces, one thing becomes immediately evident: the environment or the physical world as such cannot be put in a box labelled 'structural history' and defined as belonging solely to a history of *la longue durée*.

Resounding events, such as a great flood, may not be surface manifestations of these larger movements. Furthermore, these sudden, 'resounding events' in nature or the man-made environment, may change the geographical conditions and establish a new structure of long-term relevance within which future short-term events may be explained. The role played by water-society relations thus cuts across the temporal dimension and the distinctions made by Braudel. It is much more dynamic and complex than a 'dead sea'; it belongs both to the *longue durée* and to short-term events (sudden floods or droughts or a morning rainfall or a changed river caused by a dam across it). The water-system approach thus avoids the fallacy of relegating nature to an introductory chapter, as a geographical, almost permanent and unchanging backdrop to human action. Research and reconstruction of what are here called hydrosocial disturbances will deepen our understanding of temporalities in general.

A focus on water rather than on geography as such, always conceived of or structured in certain ways, does not entail a particular adherence to any one of these three timescales. The local hydrological cycle and its variations and the role of large river systems do not imply a social reality of a certain duration, or of the long term, in spite of their permanence, since they can undergo and cause or stimulate dramatic short-term changes that lead to social adaptation to a new condition under which other relations between water and society structure social and economic life. The atmosphere traps the moisture of the earth, so the amount of water on the planet is more or less constant, and the water we drink is the water that the dinosaurs drank. There is no such thing as 'new' water, and it is impossible to describe water as undergoing a 'birth, youth, mature period and death', a much-used metaphor for historical periodisation and for life itself. But since the water that falls as rain today is the same water that fell on prehistoric man, the dinosaurs and the earliest forms of life on earth, nature is engaged in a never-ending process of recycling our planet's water. Rainfall lands on the earth, heat evaporates the water, the vapour rises and cools and forms clouds, and once again the rain falls. Water's time is therefore both cyclical and eternal. There is no new water to create, and there is no old water to lose, and therefore the relationships of societies with this resource are very different from their relationships with other controllable resources. The continuities and changes in how humanity has related to the planet's water are therefore some of the fundamental and permanent conditions of world history. The reconstruction of these interactions will therefore be a history of deep rhythms but also, and very importantly, of sudden events that, if water is put in the centre of the picture, will open up for new periodisations of world history.

However, the enormous variety in society/water confluences can also illustrate Braudel's useful notion of simultaneous times, but from a new angle. The hydrological cycle and the structuring roles of rainfall and river patterns make the combined presence of the past, present and future of water an important aspect of social life. The current way of managing water reflects both the physical and anthropogenic layers of waterscapes of the past as well as expressing the dominant current ideologies surrounding water. As substantiated, efforts at controlling, moving or channelling water will have long-term effects, because of its importance to social life, its ability to stimulate adaptation and the long-term structural implications of dams and canals for economic activities, patterns of settlement, etc. Analyses of the relations between the hydrological cycle and the hydrosocial cycle and the study of the role of water in societies will evade the sense of material necessity often associated with Braudel's notions, simply also because dramatic water events may change the structure of the *longue durée*. During this long history of society-water interactions there are many examples of individual water engineers fundamentally altering the basic structuring properties of a river system. Large dams, for instance, have completely evened out seasonal variations in the very flow regimes, to which societies in their pre-dam existence had adjusted their rhythms. Such structures fundamentally alter the lives of both people and ecosystems downstream 'forever', for several direct and indirect reasons. A single event can thus disturb the regularity and predictability of everyday life, with profound long-term impacts that establish new forms of regularity and predictability to which societies adjust.[29]

A focus on water-society relations therefore makes the distinctions between geographical, social and individual time more complex. Extreme natural hydrological events and societies' ability to affect river systems fundamentally by sudden water control initiatives may have long-term, potentially irreversible impacts in both the short and the long term. Both the social and the physical water world can be impacted in both the long and the short term, even potentially creating brand-new structures of importance to how people relate to their waterscapes. This type of simultaneous duality sheds new light on the question of temporality. A water-system perspective permits the simultaneous integration of the *longue durée* and the short term. The latter is not regarded within this way of thinking as solely situated within the longer duration that envelopes it, but, in contrast to Braudel's ideas, it demonstrates its repetitiveness and highlights the logic of the dramatic event but occurring within a structure; it pinpoints the position of the event along the cyclical movements of conjectures but related to a geographical structure of the long term. Dramatic short-term changes in the waterscape may represent a break with the past and should not always be explained as being embedded in it,[30] and hydro-historical reseach, yet in its infancy, will make that clear. The way a sudden flash flood caused by sudden changes in atmospheric pressure and a course change caused by silt being deposited over centuries in the same river at the same time will be located in different temporalities, and these changes are not accounted for by the *longue durée*. The continuing dialectic relationship between human activities and natural phenomena manifests itself in both the waterscape and society at definite periods in history. Where the Braudelian tradition encounters problems in explaining transformation and change,[31] a focus on human adaptations and waterscape brings such issues to the fore. A dynamic vision of history and geographical time highlights both continuity and the *longue durée*, and cyclical change and short-term political and economic developments.

A historical-geographical archaeology of water-society relations will challenge the idea that there is an irreconcilable divergence between natural and historical time. It is precisely due to this fact – the monsoon brings rain, water always evaporates in the heat, rivers swell with snowmelt in the spring, etc. – that the cyclical and directional nature of water exists independently of human concepts of it at the same time as it influences social and human perceptions and changes of time. Water's elusive imprint is a permanent, though continuously changing feature in nature, society and in the life of every human being in the past, present and future.

Based on an acknowledgement of the role of interactions between society and water over time, the following notions can be suggested: one timescale must be geographical and climatic, a history in slow motion in which permanent structures in the waterscape and this relationship can be detected and highlighted. Unlike Braudel's perspective, however, this approach does not play down the ordinary sequence of political, military and societal events, which are not reduced to something almost insignificant compared to the long geographical cycles of imperceptible, slow, repetitious, endless movement. This difference is related to different conceptions of time that the water issue as discussed here can illuminate.

Time is a very important factor in understanding the types of action and interaction that take place between dynamic water structures and dynamic social systems. By regarding this relationship in different time perspectives it becomes clear that the water-humanity relationship is not one of direct determinism. Since water is universal and has always been a vital though changing and changeable resource for every society, the issue of how to understand duration, time and discontinuity becomes crucial, and a focus on water will deepen our understanding of the role of time in social development.

WATER-SOCIETY INTERACTIONS AND STRUCTURATION THEORY

Anthony Giddens' famous structuration theory overlooks the importance and impact of the water-society nexus and thus cannot be accepted as a universal theory of the duality of structures. It suggests the balancing of agency and structure by referring to the *duality of structure*, but the structuration theory misses the heterogeneous complexity of real time–space distanciation and the fact that routine actions take place within or in relation to different natural systems that may also be radically shaped by human actions, whether instantaneously or over time. Giddens' theory was an alternative to the orthodox consensus in sociology that regarded behaviour as the result of forces that actors neither controlled nor understood. He argued instead that social action creates these structures, and that structures exist only in and via the activities of human agents (Giddens 1989: 256). The theory emphasises that action is conditioned by existing cultural structures and creates those structures through the process of enactment; they are products of regularities of social reproduction (Giddens and Pierson 1998). Structures are defined as 'rules and resources organised as properties of social systems'. Human social activities 'are recursive, not brought into being by social actors but continually recreated by them via the means whereby they express themselves *as* actors. In and through their activities agents reproduce the conditions that make these activities possible' (Giddens 1984: 2). Structure consists of rules and resources recursively implicated in social reproduction. Giddens' structuration theory is therefore clearly embedded within the dominant social science tradition where what explain social facts as a structure are other social facts. Giddens conceptualises the social world – what people are saying, doing and believing and their actions and interactions – as something that can be understood as being fundamentally unconnected to or uninfluenced by the natural, physical world,[32] and that causes problems in analysing real time–space distanciation. According to this theory, structures enjoy an ontological status that privileges them over agency, and they thus determine material and symbolic relationships of myth or language. The social systems that emerge, consisting of reproduced structures, should therefore be analysed as the result of social factors, all in line with Durkheim's dictum.

Giddens' theory of society and structures therefore takes no account of where in space social life and actions are reproduced, and whether the recurring patterns of social actions occur within, for example, to take some extremes, a desert environment, in settlements in a marshy region prone to repeated flooding or in cities and villages where the pulse of the annual monsoon defines the rhythm of life. The modality of a structural system enabling human action is therefore not only the product of translated actions. The patterns of translated actions over time will reflect different relationships to different waterscapes, and because the hydrological cycle or waterscapes in some cases may even directly impact on social structures (for example in the case of sudden catastrophic floods) it may also occur irrespective of translated actions. The theory overlooks the fundamental diversifying and structuring role of the hydrological cycle and the workings of water. On a grand scale these create climatic zones and diversify landscape formations that will impact on patterns of settlement, economic activities and trading patterns. They create diverse structures of seasonality, precipitation patterns and trading routes that in turn systematically influence human actions and agents, and thus also those structures created by human action. More locally and in a short time perspective, whole communities may be swept away by floods or droughts and social institutions changed or reformed in their wake.

Structures are created by action, but we cannot understand similarities and differences in the acts of the actors without bringing into the picture *a priori* physical or hydrological structures that are also changeable by both nature and by the actors themselves. For example, the structural impact of the Nile on Egypt via social action creates new structures that by far overshadow the importance of any other structure in that society. It allows the 'binding' of time–space within the Egyptian social system, and makes it possible to understand both changes and similarities in social practices existing across time and space. The Nile is not a structure that can be reduced to a structure that enters simultaneously into the constitution of the agent and social practices and 'exists' only in the generating moments of this constitution. This structure is both beyond the realm of human control and can be impacted by man; action does not create it but might impact it, and the relationship may well change over time.

The water-system approach will make it clear that the duality of structures is of a different kind from that proposed by Giddens. In this context it is essentially a process whereby structures of the hydrological cycle or the workings of water over long periods of time create a systematic foundation for specific actions that continuously enact structures and thus social systems, and these social systems in turn express this duality. This approach recognises both the hydrological and social cycles and, in examining social systems, it examines physical *structures* and human modifications of these physical structures, conceived of as interaction and action, and ideas and plans. The physical waterscapes constitute a long-term structural but also dynamic process that operates behind the backs of humans, and which furthermore shapes actors' constraints and influences choices. Over time, particular waterscapes have produced extremely diverse through fundamental patterns of social behaviour (i.e., rice paddy/potato cultivation, artificial irrigation/rain-fed agriculture, bathing rituals in Indian rivers or cleansing rituals in the deserts of Arabia, etc.). Since every society always needs water for most social activities, the structure of the water cycle helps to deliniate what social actions are possible and rational. A social system's enduring patterns of social relationships will be influenced by both the continuous and changing natures of the waterscape or river system. The triple-layer water-system approach reflects and integrates the dichotomy between non-material social structures and material water structures, both of which need to be analysed as structuring human action at any given point in time.

Giddens' concept of 'reflexive monitoring of actions' therefore takes on another meaning in this context. Reflexive monitoring is concerned with the ability to look at actions in order to judge their effectiveness in achieving their objectives. With regard to water and water control and water usage, agents in some cases reproduce structures through action, but in other cases this is impossible because the structures that frame actions are not created by these actions (i.e., the annual flood pattern of rivers or seasonal variations in evaporation and precipitation). In some cases, agents do whatever they can to erode existing structures, but they may end up strengthening other structures that contextualise actions (i.e., building dams that eventually collapse, as has happened many times in history). Waterscapes continually produce structures that fundamentally impact on the arena of agency and human action, enabling the construction of *qanats* at the feet of mountains in Oman followed by the system of using the moon and a stick to allocate water among farmers, or the basin irrigation on the banks of the Nile that required social organisation and high-level state water managers, in order to develop agriculture, and human action produces and reproduces waterscapes that will structure human action again, from large dams to urban water supply and sewage systems. Agency in relation to water can thus lead to the reproduction, continuation and transformation of society.

The character and content of this human–water relationship differ in space and time, and these differences are re-enacted as social structures every day in all societies in different ways. Since water is essential to life and societies will collapse if they lack sufficient water, the waterscape and human relationships with it are fundamental and enduring, though shifting, components of all societies. The combination of physical stability and flux makes water a dynamic force of both continuity and change. The features of physical reality and physical positions, in a river basin for instance, inform the parameters for social action, and therefore the capacity of actors to produce and reproduce a particular social world. The ways in which rivers run have established structures that made some societal adaptations possible and others less likely, both before and after waterscapes were put under human influence or control, and both when rivers behaved 'normally' and when their hydrological character changed suddenly and dramatically. Examples of this are the Huang He flood of August 1931 when 88,000 square kilometres of land was flooded and somewhere between 850,000 and 4 million people died, usually regarded as the most serious natural disaster in recorded history, river captures in India, where even major rivers have been captured by other meandering rivers, or dyke collapses, such as in the Netherlands in 1953, that led to an overhaul of the Dutch water system and a strengthening of the water boards.

The water-system methodology enables us to perform comparative research to find out how far similar types of such systems tend to encourage similar social solutions or social structures, and to what extent different social solutions reflect different structures of water-society interactions. The hypothesis is that as far as other historical conditions remain equal, major structural and structuring differences in water-society relations will be an underlying factor in explaining institutional differences, not usually as a direct cause but as a condition encouraging different development trajectories and institutional diversity. There is little doubt that physical structures in some cases encourage the establishment of specific social structures, which can radically change in a moment due to abrupt changes in the waterscapes, such as floods, or to abrupt changes caused by water control efforts. In other words, long-lasting social structures can be rapidly transformed as a result of physical transformations. Thus we cannot understand why certain social structures exist without understanding the water landscapes and the water cycles or how humans have related to them in the course of time. The structure of rivers or waterscapes in societies has conditioned the choice of technology and water control efforts and plans. Waterscapes also structure patterns of settlement. Most cities have been located on riverbanks or close to other water sources out of dire necessity, but rulers, from the Romans to socialist local governments in

Scandinavia to Stalin and Vivendi, have also made water come to the settlements and cities, but always in ways that are constrained by the particular waterscape, whether man-made or not. Different waterscapes and different confluences between society and water have encouraged development of different technological systems and traditions to take these different conditions into account, as they also, although in very different ways, encourage different institutional traditions (e.g., the role of public works departments) and habits of thought (e.g., feminine attributes of Asian rivers and flood myths in many religions).[33] The water-system approach objects to and handles a theory of agency that reduces all acts and all structures to reflections of social meaning and action. It counters the inherent reductionism by acknowledging, for example, the need for understanding what can be called 'water structuration'.

THE 'DEATH OF NATURE' THEORY
AND THE UNCONTROLLABLE WATER

A contemporary school of thought will claim that an approach focusing on the relationship between the water cycle and societies is anachronistic, even pre-modern perhaps, because modern society or capitalism killed off nature; now nature is 'dead' and the hydrological cycle has become a hydrosocial cycle. This section discusses and criticises the ideas of the 'Death of Nature' argument and its logical consequence; the idea that the hydrological cycle is not natural any more and should be conceived of as a 'hydrosocial cycle', the aim being to further explain in another context and from another angle the analytical approach and conceptual tools suggested in this book.

There exists a very powerful historical narrative that holds that nature is dead and that 'nature does not exist any more' because of human interference with it during the Industrial Era. The dominant story goes like this: Man's attempts to control nature, at least on a grand scale, started with capitalism and modernism, unleashing forces of control and destruction. Money, greed, technology and organising capacity launched a process that still haunts humankind. This idea of the onslaught of modernity on nature forms the premises of much current research on human–environment relationships. This story captures important developments and reflects the fact that the untouched wilderness is a thing of the past. Humanity has left its footprints all over the globe and there is very little 'pure' nature left, resources have been depleted, and even climate has been affected. But when we consider the relationship between water and society, then the statement that nature is dead is not only unduly anthropocentric, it also contradicts undeniable facts.

First, we cannot talk about 'the situation of nature' in general without talking about water, since no natural resource is more plentiful than water. Arguments that suggest that nature is dead must therefore also mean that 'water is dead'. And conversely, if water is not dead, nor can nature be. Secondly, if the argument is that nature is dead due to human interference then water should definitely be dead since humans have attempted to control and tame this part of nature for a longer time and more systematically than any other part of nature.

Water has been regarded both instrumentally and as a religious, ritual object from the time of the first irrigation canals of ancient Sumeria and Egypt until the present. The canal builders of Mesopotamia, the land between the rivers as the name signifies, had their names inscribed due to their importance for bringing water to the farms and to the city states, while it was the ejaculations of the river god, Ea, that created the world and life. In Egypt the first representation of the first Pharaoh shows him excavating a canal, and thousands of years ago the Egyptians dug a canal from the Nile to the Red Sea, many kilometres across a rainless desert while they worshipped Isis and Osiris and the other gods who presided over the life-giving waters. In China the Jade Emperor Yu became a legend because he tamed the wide, marshy river deltas, thereby creating China and thus the world, while almost 2,000 years ago China started to construct the Grand Canal from Yangtze in the south to Beijing 1,800 kilometres to the north, an engineering project that in complexity and control surpasses many projects in today's world. The Roman emperors radically altered the local waterscape and brought water into the imperial city via aqueducts and underground tunnels from reservoirs outside the city. The Mayas built their reservoirs to store rainwater more than 1,000 years ago, and the rulers of the Mayas were also gods because they were believed to be the guarantors of the water and the water-control system. Water was also controlled in Sri Lanka from the twelfth century and in Yemen where the Dam of Ma'rib dammed the Wadi Adhanah almost 3,000 years ago. Water has thus been controlled, tamed, dammed and piped for thousands of years. So a view of history that limits attempts to control, subdue and tame nature to the period of modernism and capitalism is mistaken. People have long managed to control nature on a grand scale, modifying entire river systems, As far as the most central of all resources in society are concerned, no radical changes in water modifying technologies are connected to the Industrial Revolution or capitalism, except that in some parts of the world the level of technology that made a more intensive exploitation of the energy of water possible was higher, while the fundamental relationships remained the same. During the 1990s, historians and social scientists published analyses about the 'death of rivers'. The Colorado and Columbia Rivers in the USA became cases in point,[34] 'raised, fattened, and slowed', they were dead rivers reduced to pure technology. Columbia River had been transformed to an 'electricity-irrigation-transportation machine'. 'There are no rapids at all, nor are there waterfalls, riffles, eddies, sinkholes, or a single "agitated gut": The Columbia does not flow, it is operated.' These analyses capture a clear historical trend in the twentieth century; due to the importance of water and the growing multifunctionality of water, many rivers running through societies became engineered rivers.

The Colorado River is definitely one of the most controlled and modified large river systems in the world, with the Hoover Dam as its foremost symbol. But to talk about the 'death of nature' in connection with the Colorado River is not very appropriate, because it neglects the issue of scale in the water cycle. Water – unlike other resources that are exploited by societies – is never totally 'killed off' or destroyed. It exists in both nature and in society in the same form and it always re-emerges as clean water as it falls from the clouds – whether on the surface of watersheds in the jungle or on the roofs of skyscrapers in cities. Unlike fossil fuels, forestry, phosphate and other natural resources, the total amount of water does not diminish even though humanity is consuming more and more of it, since there is as much of it now as there was when societies were first established. Moreover, water cannot be entirely controlled, since eventually it always slips out of the hands of the controller, much of it literally evaporating. And, most fundamentally, the governing forces of the hydrological cycle, first and foremost solar radiation and then the wind, are beyond human control. The Colorado River can be controlled by the US Government, so the river with its naturally produced banks and waterfalls may not exist any more, but the waters in the ditch are not dead. The water re-emerges somewhere else, still beyond the full control of mankind. The 2014 drought, which at the time of writing this book is dealing a tough hand to the arid regions of North America, illustrates just how the annual snowfall in the Rocky Mountains – or lack of it – is ultimately establishing the limitations and possibilities for managing and distributing water shortages in Arizona, Nevada and Mexico in the coming years because it is the major force deciding how the water is running in the Colorado.

There can, however, be no doubt that the primary meanings ascribed to water in the twentieth century were produced by those controlling water through engineering, law and economics. Flows of water in many countries in the world thus embody more and more the values and priorities of modern societies and technology as well as 'social struggles and conflicts' (Swyngedouw 2004: 4). But, most importantly, the workings of the water cycle and the enormous amount of water that is not part of this cycle make it evident that the earth's water system is still only marginally influenced by human activities, although more and more of the water that falls as rain and as run-off is controlled by humans in one way or another. Although most rivers in the Western hemisphere and in China are controlled and tamed and exploited, it seems odd to generalise this situation for the world as a whole. Ethiopia, the potential water power centre of East Africa and the Middle East, had until recently, with the building of the Renaissance Dam, barely tamed any of its major rivers. Nepal and Bhutan have gradually begun to dam some of their rivers. In a world in which enormous river basins, like the Congo in Africa, which is scarcely managed and where seasonal streams run freely as run-off down mountainsides and in the bottom of remote mountain valleys and millions of mountain springs emerge under no form of control at all, it is at best premature to talk about the 'death of rivers', 'death of water' and hence also 'death of nature'.

A logical implication of the concept maintained by the 'nature is dead' theory is that it is no longer fruitful to use the term 'hydrological cycle'. It should rather be substituted by the term 'hydrosocial cycle', because 'practically every body of water on the planet bears traces of human involvement in the form of minute quantities of anthropogenic substances such as chlorinated organic compounds' (Linton 2010: 229). The argument is that everywhere water bears the traces of its social 'background'. The water question is within this perspective not simply 'about water', since water is a 'process' that 'occurs through us', a 'self-identical object', and an identity formed within social relationships. A typical expression within this way of thinking will be that water becomes what it is in accordance with a particular kind of engagement, that is, water becomes what we make of it; it is basically a social construct.

Of course, the water-society cycle is a social recycling process, but it is also impacted by physical processes existing outside the social world, and whether it is conceived by humans or not. New research in the history of the hydrological cycle makes this philosophical argument empirically more and more evident and relevant. The water cycle is still primarily a physical process that to a certain extent and in some areas is influenced by social relationships and is conceived of and understood in different ways by different social actors. To argue that water is what we make of it is unfruitful and empirically absurd, since water exists without humans at all, and existed long before societies were established. First, the hydrological cycle has not therefore *become* social although there are human elements to it, or humans influence or humans conceive of it. Since the cycle is a cycle that is part of the earth system and water re-emerges as clean water, untouched by humans, we cannot describe it as having become social. The very nature of water circulation must not solely or primarily be described in social terms. To do this would be to overstate the power of human beings. Compared to solar radiation and the force of the winds governed by water's movement, the power of man is negligible. And secondly, to argue that the hydrological cycle has *become* social requires an answer to the questions: just when did it change from being a hydrological cycle to a hydrosocial cycle? What are the variables that caused or represent this extremely important shift in history? And, finally, how can this great shift be described, that is, how exactly has the hydrological cycle been changed and where and to what extent? The proposition cannot be substantiated. To argue that the waters of the oceans, the waters of the Amazon that represent some 25 per cent of all river water in the world or the waters of Lake Baikal, 25 million years old, 1,700 metres deep and containing 20 per cent of the world's total, unfrozen, freshwater reserve have become socialised or are 'a process that occurs through us' is wrong from whatever perspective one regards it. We can say that a lake or a river is destroyed, that forests are destroyed, that oil resources and fish stocks are depleted and have become victims of social process, but it is not possible to say the same about water.

The term 'hydrosocial cycle' has, however, important ontological and epistemological implications. It also overlooks an important characteristic of water, and one that theoretically undermines the 'nature is dead' theory or the concept of the 'hydrosocial cycle'. Water is unique since it does not change its character by being appropriated by society or humans and will always ultimately escape the attempts of mankind to take control over it, since it is always on the move, from one place to another and from one physical state to another (from liquid to vapour to liquid). Thus, by its very character and relationship to humanity, it undermines the conventional boundary between society and nature, with consequences for how we understand nature–society distinctions and relationships in general. The waters in a river may be polluted, or a wetland be destroyed, but the water itself will always reappear, after its 'journey' through society as a social good, as clean natural water. Unlike an apple, which stops being an apple after it has been eaten, or a tree that ceases to be a tree after it has been turned into timber, water is always water, in the environment and even after it has been consumed by a biological organism. For the same reasons, and unlike other resources, it cannot be completely appropriated. The implications of these insights are huge for how society, modernity and history as such are viewed.

What our efforts to control water have demonstrated is that it cannot be completely controlled, and that it always re-emerges as pure nature. The efforts of societies to bring it under social control demonstrate the limits of socialisation. The temporary nature of social power is contrasted and highlighted by the permanence of the hydrological cycle. It means that the water-system approach enables us to study what happens when natural and anthropocentric forces combine, and when our activity reinforces trends in the landscape and the opposite. Instead of talking about only the hydrosocial cycle, one should analyse both the hydrological and the hydrosocial cycles and their inter-relations and the historical development of hydrosocial redistribution. It has proven to be extremely difficult to disentangle the relationship between nature and society, but the problem should not be solved simply by getting rid of it through definitional exercises, since the water cycle will continue to impact on social and geopolitical developments in the future.

THE 'UNITY OF NATURE' AND ITS DECONSTRUCTION

This book argues in favour of a binary research strategy focusing on water, not as a one-factor explanation but rather as an effort to battle reductionism and determinism: (a) to propose a water-system approach to the empirical study of development and history because of the crucial role played by water on our planet and in social life, and because such an approach will help to solve some important conceptual and methodological problems of research on societies in general; and (b) to dismantle the way in which established dichotomies between nature and society have been drawn, and deconstruct the notion of nature as a useful *analytical* category, not least because this would make it easier to theorise about extremely multifaceted and complex environment–society relationships.

One reason is practical and empirical; it is not possible to examine empirically the relationships between and interdependences of societies and nature or environment as a whole, since both nature and society are such extensive, complex phenomena. The terms 'nature' and 'environment' are such elastic terms and cover such a myriad of variables and aspects of importance to societies that meaningful, empirically oriented research or well-focused discussions become virtually impossible. For the purpose of the argument here it is not necessary to disentangle all the various threads of discourse surrounding the concepts,[35] which can refer to anything from the object of study of the natural and biological sciences, to anything 'non-societal', as well as metaphysical issues regarding different modes of being between the human and the natural. As Raymond Williams noted: nature was 'perhaps the most complex word in the language' (Williams 1980: 221).

Moreover, the conceptual and methodological implications of the fact that the relationship between society and nature will vary according to what kind of element in the non-human sphere one is emphasising have not yet been adequately addressed. The deconstruction of the 'nature' concept here has therefore nothing in common with the constructionists'

232

discussion of 'natures'. They argue that there is no single 'nature', only 'natures', and that 'these natures are not inherent in the physical world' but rather 'discursively constructed through economic, political and cultural processes' (Macnaghten and Urry 1988: 95). The argument of this book is the very opposite: because nature is a physical reality, and because it is so diverse and enters into so many relationships with societies, and that environment without a referent can mean everything that surrounds everything that exists, it is unwise to study the evolution of human–nature relationships or society–environment relationships in general.

Comparative studies of human–nature relationships or environment and socio-cultural relationships have proved to be problematic due to the fact that different elements of nature have played various roles in different societies and in relation to different societal sectors at different times.[36] Some of the best-known studies of environmental history and human–environment archaeology have therefore in practice not analysed the relationships of societies to the environment and their interconnectedness with nature as a whole, but only certain elements in nature and/or sub-systems of societies' relation to them.[37] The historiographical tradition demonstrates that some sort of deconstruction of nature as a single entity is in practice unavoidable in empirical studies.

The criticism of the term 'nature' should be qualified. To reconstruct how societies' *ideas* or *discourses* about nature have revolved around discourses about nature conceived of as an entity is important.[38] The concept of 'nature' has been fundamental to the ways in which science perceives and conducts itself. It has generally been understood as a unity and as an aggregate opposite in relation to society. This dichotomy between nature and society has a long tradition and is deeply ingrained in Western thought (Glacken 1967), traced back at least to Aristotle for whom nature (*physis*) was that which is not made by humans, in contrast to *techné*, that which is of human origin. This tradition in philosophy and religious speculation and discourse has been concerned with nature as a whole – often and for a long time only as a God-given entity – and the place of humanity within it. Romantic and pastoral attitudes have envisioned nature as being opposed to society – as a refuge. As a holistic 'other' it has been regarded as offering physical, psychological and spiritual respite from the fragmented lives of modern society. The literature on the history of perceptions and images of nature and the environment is extensive (Glacken 1967; Worster 1985a [1977]; Pepper 1996; Coates 1998; Buttimer and Wallin [eds] 1999), and it all testifies to the importance of this term. More recently it has also become a core term in the influential normative and ideological project of the 'saving of our planet', and has as such influenced research designs and research questions in fundamental ways. Within this movement, the unity of nature or environment is a metanarrative, and the term 'nature' has therefore, and is intended to have, a useful and sought-after totalising function. Historical studies of environmental philosophy and thinking have mushroomed during the past few decades (e.g., Pratt et al. 2000; Jamieson 2001; Gottlieb 2003; Selin and Kalland 2003; Turner 2005; Callicott and Palmer 2005; Radkau 2008). The narrative aim of many of these studies has been to foster a holistic approach to nature and the environment, supporting the modern environmentalist movement that has preached and developed a philosophy of human conduct related to 'nature' as a whole. The goal of learning to see nature as a unity comes out very clearly: 'One may hope that research in environmental philosophy will eventually produce a single set of universally valid environmental attitudes and frameworks that will transcend all [...]' (Callicott and Ames 1989). This view on modern ecology as a way 'to healing our planet' indicates the importance of it; it has been described as marking off 'a particular era in history' (Worster 1985a [1977]: 360); the 'Age of Ecology' (Worster 1985a [1977]: 342). In 'deep ecology' ecology is by definition holistic and even self-healing as an organism (Rolston 1992).

The suggestion of a deconstruction of nature for analytical purposes might meet politically and ideologically inspired opposition, since, as Worster has suggested, Gaia is the most widely accepted scientific metaphor in the Age of Ecology, and the main function of the Gaia metaphor is precisely to emphasise that nature should be regarded as a single organism.

> Gaia theory looks at the whole Earth from the outside and sees it as a live entity; modern science looks at the surface details and gives us an inventory of the parts [...] There is more to Gaia theory than a change in viewpoint; the theory enters the realm of emergent phenomena; a place where the whole is always more than the sum of the parts. (Lovelock 2000: xii)

The Gaia thesis and the new ecological movement inspired a veritable flood of literature that argued against what was called the 'atomistic-mechanical' image of nature. The failure to see nature as a unity and as an organism has been condemned as the root of the ecological crisis. The holistic-organic reality 'discovered by contemporary ecology' represented the way out (Callicott and Ames 1989), and has been described and hailed as crucial in order to save our planet. The monumental five-volume work *Environmentalism: Critical Concepts* (Pepper et al., 2003) is telling in this context: it discusses a number of critical concepts, but not the fundamental constituting idea of nature or environment as *one thing*.

Nature, through this lens, must be seen as a unity,[39] or even as a single organism. Studies within this tradition typically argue normatively in favour of the nature concept; science 'can be a corrective to the prevalent tendency of humans to see themselves as separate from nature, above nature, and in charge of nature'.[40] The goal of examining the reciprocal relationship between humankind and nature has been described by some as forging a 'New Ecological Paradigm'.[41] This novel social science environmentalism has often aimed at fostering a holistic approach to viewing and studying nature and the environment. Accordingly, humans and nature should be regarded as an inclusive whole, primarily for political reasons; it will lay the foundation for political success, such as 'green change',[42] or more 'green politics'.[43]

As has been shown in this book: water is the one identifiable part of nature that all humans relate to in specific ways and it is theoretically possible and methodologically fruitful to single out water as an aspect of nature for analytical purposes. It is possible to reconstruct, describe and understand its movement and role in nature and in society, and thus evade the problems created both by natural or biological determinism and social reductionism and constructionism. I have suggested an analytical approach and some concepts that have been tested in relation to some central empirical questions in history and social development. To understand the importance of water for human and social evolution and diversity is, however, a long-term task, an unending series of confrontations between explanatory efforts and the hard, pitiless facts of history and empirical social science. This book is just a stepping stone to a future where, with the help of new technology and more data, it will be possible to push the approach to a gradually more complete systematisation of many more combinations in an almost endless number of empirical schemes.

EPILOGUE

'An Unstable Foundation of Running Water'

When the Mars *Pathfinder* probe touched down on what was described as an ancient flood plain, it was confirmed that there had been widespread flowing water on the Red Planet in the very distant past. Some scientists identified landscapes which they believe were carved by torrents of water with 10,000 times the force of the Mississippi River. Geological features and outflow channels evidenced that billions of years ago liquid water flowed on the surface of Mars. The year after, in 1998, NASA confirmed that Lunar Protector had found evidence of some 300 million tonnes of water on the surface of the Moon. These frozen reserves may be space travel's ticket to the universe, but still they are only a minuscule fraction of the earth's water resources. The Mars *Curiosity* rover has identified since it touched down inside Gale Crater in 2012 what scientists say is a river that ran on the planet for a long period of time, and the surface soil on Mars contains about 2 per cent water by weight.

The more we learn about our solar system the clearer it becomes that what distinguishes our planet is not simply that there is water here, but that it is a lot of water and that it moves in the atmosphere and runs across the globe and through the landscapes in ways that have made evolution and social life possible. The Earth, the planet's name, becomes more and more imprecise, and the waterblind perspectives on human development, on the formation of societies and social life and development trajectories become more and more archaic.

The German philosopher Friedrich Nietzsche described the world as 'an unstable foundation of running water',[1] a useful expression for underlining the futility of identifying fixed truths. But, paradoxically and as this book has shown, water is definitely the foundation for social life and its fluxes have beyond doubt fundamentally impacted on development and social diversity. Research has shown that our world has developed on the eternal but at the same time variable and unstable foundation of water in flux, because liquid water is not a bit player in the theatre of life but a headline act.

GLOSSARY

Below is a list of terms used and defined in this book.

Age of water insecurity: The period in world history that started a short while ago but will last for ever; it is an era in history where the insecurity about the future behaviour and flux of water – whether we live in the age of floods or in the age of droughts, whether the sea level will rise or not, whether the glaciers will disappear or not, etc. – will fundamentally influence public discourse and political and economic considerations and decisions.[1]

Agro-water variability: This recognizes water variations as a factor of fundamental importance to the development of different food-producing regimes, at any given time and at any given place, creating possibilities and limitations that through human ingenuity have given birth to a wide variety of water management practices, rural and food adaptations and societal organisations.

Confluences between water and society: The fluctuating, ever-lasting and manifold meeting points between water and society on all levels, reflecting and impacting in multi-directional ways both nature and societies. Such confluences are a universal trait of societies, and thus all societies and their long-term and short-term development can be fruitfully analysed by focusing on them.

Eurasian raincoast states: States whose long-term history and patterns of economic and social development have been strongly influenced by being located on the raincoast of north-western Eurasia – a geographical area in north-west Europe characterised not only by abundant rainfall but year-round precipitation and modest evaporation.

Historical-geographical archaeology of water-society relations: A research strategy that aims to reconstruct past development of the role of water in society and the role of society on water in a geographically defined space and over long timespans.

Hydraulic calculations: Assessments of a water-society system's production capacity in a broad sense, including the recognition of its physical and hydrological capacities as well as its capacity to produce socio-economic results; it is an inclusive term formulated within an understanding of water management as an interdisciplinary subject.

Hydraulic design: Water management plans aimed at controlling and using water that will always reflect existing waterscapes and water-society interactions and relations, and that will always impact on existing waterscapes and societies as well as the interfaces between the two.

Hydraulic state: A state whose economy and geopolitical position is to a large extent based upon extensive control of water in some form or another. The term is thus here used in a purely descriptive manner, not carrying those normative connotations of despotism and autocracy that in general are connected to it.

Hydro-historical approach: A cross-disciplinary research approach and method that aims to utilise all kinds of data to reconstruct both deep-seated continuities and slow and abrupt changes in the waterscape and in water-society relations – from traditional archaeological method and climate data of the past, palaeontological data, geological science data in addition to all kinds of written sources, etc.

Hydrological cycle: This term is used here as it is normally used in the scientific literature, but with two qualifications: it is describing the continuous movement of water on, above and below the surface of the earth, thus being clearly distinguishable from water in general (since large amounts of water in the oceans and in the ice-sheets do not move around in the hydrological cycle). It is distinguishable from the hydrosocial cycle because large amounts of water in the hydrological cycle are not impacted by the social. The term hydrological cycle does not separate water from its social context, but underlines the fact that it is a physical process in nature.

Hydrosocial cycle: The never-ending flow of water in societies, capturing how water is socially used as well as constructed, impacted on by society as well as impacting on society. The hydrosocial cycle is defined in relation to the concept of the hydrological cycle, although it does not modify the term, but complements it.

Hydrosocial disturbance: A term that captures either changes in the hydrological cycle or in the hydrosocial cycle that significantly impact on the relationship between the hydrological and hydrosocial cycle.

Hydrosocial redistribution: A term that captures either changes in the hydrological cycle or in the hydrosocial cycle that significantly alter the relationship between the hydrological and hydrosocial cycle.

Monsoon states: States whose economy and social organisation are strongly influenced by the irregular regularity of the monsoon.

Scandinavian raincoast: The coast of western Norway, characterised by abundant rainfall and all-year precipitation (an area with an annual mean of above 2,000 millimetres per year and with more than 200 rainy days per year).

Social cycle of water: The never-ending flow of water in societies, capturing how water is impacted on by society and impacts on society.

Water cycle: The product or the result of a non-dialectical relationship between the hydrological and hydrosocial cycle over time. The water cycle should thus be understood in connection with the hydrological cycle and the hydrosocial cycle and be seen as a long-term, socio-natural process by which water and society make and remake each other over space and time. The workings of the water cycle can be reconstructed, and its study will offer insights into both natural processes, how human activity impacts the hydrological cycle, as well as the social construction and production of water with all their social implications.

Water imperialism: A kind of imperialism whose expansionism is encouraged and stimulated by the prospect of controlling water for economic, military or political reasons.

Water management: An interdisciplinary activity which needs to be understood or defined not in terms of engineering alone or in terms of sociology alone, but in terms of its engineering, historical, economic, sociological, cultural and ecological dimensions.

Water Planet: A term used instead of the term 'the Blue Planet', since the latter focuses on colour while the first focuses on what makes life and social life possible on the planet. The 'water planet' term reflects and captures the importance of the enormous and greyish-coloured cloud-systems and ice-sheets that form crucial parts of the water planet, and is not concerned with the reflection of light.

Water structuration: An analytical term that can capture and analyse how waterscapes and water systems structure social production and reproduction and how different water-society systems make different actions and development trajectories likely or possible. This term is used in hydrological sciences in a very different way.

Water-society studies: Studies about the relationship between water and society, underlining that this research topic is a potential sub-theme of all disciplines and that it also flows across disciplinary lines thus continuously asking for new approaches and collaborative engagements between disciplines.

Water-society systems: A descriptive term of specific water-society relations that are so complex though so stable that they can be seen as kinds of enduring adaptive systems characterised by cycles and uncertainty, and where social systems and waterscapes are seen as fundamentally coupled and co-evolving.

Water-system approach: An analytical approach underlining the importance of studying water-society relations in a more systemic way, concerned as much with ideas, institutions, technologies and economic aspects as with natural aspects of the waterscape, at the same time as it includes a loose-knit web of analytical and conceptual propositions, some more central than others and some spun more tightly than others.

Waterblindness: A world-view and a view on society that analyse social development and fundamental historical processes as if they are unconnected to and uninfluenced by water and how it is controlled, harnessed or adapted to.

Waterscape: A water landscape created by nature or by human modification or by both. The term as used here underlines that it should be seen as something more than a cultural landscape, although it can also be a cultured waterscape. It also underlines that it should be seen as something more than a natural water landscape, although it can also be a natural water landscape.

Waterzone: A broad water-geographic division, based on distributional differences of precipitation and run-off characteristics that have a fundamental importance for biodiversity, flora and fauna and hence for the potential for rural organisation, and covering small or large areas depending on the characteristics of the waterscape of differential importance for societal development.

NOTES

CHAPTER 1: THE NEED FOR A PARADIGM SHIFT

1   The image was taken by Bill Anders, one of the astronauts on Apollo 8, through the window of the spacecraft.
2   This was demonstrated when Saudi Arabia in the 1990s discussed plans to tow icebergs from Newfoundland to the Arabian Desert, and again when the project of towing water from Turkey to Cyprus in giant plastic bags was launched.
3   Diogenes Laertius' more than 2,000-year-old story about the Greek philosopher Heraclitus is telling. Heraclitus, remembered first and foremost for his theories about water being always on the move and that it is therefore not possible to step twice into the same river, died because of his misunderstanding of the role of water in the human body. He died lying under a baking sun, buried in dung, fighting what he thought was the dangerous and unhealthy water in his body. He saw the world within the dichotomy of fire and water, and his aim was to become 'dry', because the 'dry soul' was the wisest and best soul, and for souls it is 'death to become water'. He even 'asked the doctors if one could, by emptying the intestines, make water pour out. However, when they said this was impossible, he stretched himself out in the sun and ordered boys to plaster him over with dung. He stayed there, stretched out, and the next day died and was buried in the agora (DL 9.3–4, quoted in Chitwood 2004: 79). Heraclitus was obsessed with moisture but obviously did not know that the water balance in the body is crucial for survival. In fact, it was quite recently that this fact was confirmed beyond doubt.
4   It is said about John Locke in the introduction to the 1824 edition of *An Essay Concerning Human Understanding* that one of his most peculiar traits was that he only drank water and that he drank a lot of it because he thought it was very healthy (p. 30). Well into the twentieth century, the crucial importance of water and the amount of water in our body was not appreciated. When, in 1936, Arthur Shipley in his book *Life* (first published in 1925) wrote 'Even the Archbishop of Canterbury comprises 59 per cent. of water' (p. 18), this knowledge was about to be common knowledge among people who studied the body.
5   Also the body temperature is related to water, since it is determined by two parameters: heat gain and heat loss. Heat gain is basically linked to heat production and it is fairly constant. It is related to the basal metabolic rate and increases during exercise and during the process of shivering. Heat loss is related to radiation, convection and conduction from the external surfaces (Milton 1998: 4).

Body heat is lost by conduction to air and other objects in contact with skin. The kinetic energy of the molecular motion of the skin is determined by temperature; energy is transferred to the molecules in contact with the skin if the objects are colder than the skin [...] Water has a high thermal conductivity (0.561 W/m·K), more than 20 times that of air, and a high heat capacity; thus, heat loss is much higher in water than in air (Morimoto 1998: 86).

To illuminate this point, one may use the sauna as an example. The temperature in a sauna is normally between 65 and 100 degrees Celsius, but the importance regarding the felt experience is the humidity. In the beginning during a sauna the heat is dry but it turns damp as soon as water is thrown on the hot stones. Throwing water on the sauna stove produces steam and makes the air feel hotter; the more water that is thrown on the hot stones the hotter it feels. However, it is only the humidity which changes, not the temperature (Leimu 2002: 72). 'Water has a high heat vaporization, and it is the only mechanism available for the reduction of body temperature when the environmental temperature is higher than that of the body surface' (Morimoto 1998: 80), and one starts sweating. Turning to a coldwater environment, shivering is the opposite bodily function of sweating.

The important factor is that 'The thermal properties of water differ from those of air by a 20-fold higher heat conductivity and a specific heat about 1000 times greater than air. The average skin temperature in a water element will, therefore, be very close to the water temperature' (Nielsen 1998: 133). Thus, the skin temperature of a person without clothes in heavy rain will soon be the same as the temperature of the rain. The colder the water is, the more it affects not only the skin temperature but also the core temperature of the body.

During rest, a fat subject (36 per cent body weight fat) can maintain heat balance in water 10°C lower than core temperature, while a lean person (less than 10 per cent of body weight in fat) can tolerate a water temperature no more than 3.8°C lower than his body temperature without shivering during rest. In other words, the prescriptive zone in water is very narrow – only about 5°C (Nielsen 1998: 134).

6 Useful studies of aspects of the history of dam architecture, hydrology and engineering have of course been published but they have not been interested in how developments have been framed and impacted by the three layers and their interconnections.

7 Two examples from the two richest countries in the year this book was written: in New York in 2013 the mayor decided to invest about $20 billion to protect the low-lying Manhattan district from climate change and further billions to ensure that the metropolis gets all the water it needs; the world's biggest engineering project ever started in China in 2002, the so-called 'South–north water transfer project'. The Chinese Government aimed to save Beijing as a capital by building a huge canal that brings water from the Yangtze River, 1,800 kilometres to the south of the city. The project should also save a large number of other big cities, replenish the water table on the North China plain and save the Yellow River from over-exploitation. By 2014, about $80 billion had been spent. But any modified waterscape, including the Chinese canals, will reflect its physical underpinnings and this might again be changed by both physical and man-made changes in the way water runs. The point is that in the long term all hydraulic structures are vulnerable to climate change. In fact, one of the great questions about the future of China is the future of this project if the climate change and the discharges of the Yangtze decrease significantly. If that happens, China may well disintegrate, as the Chinese Empire could not withstand the grave consequences of changing climate and water discharge of the Yellow River and the Great Emperor's Canal in the late eighteenth century.

8 The approach suggested here may also for the sake of clarity be compared to Lefebvre's call for a 'unitary theory'. Lefebvre argued for the need to bring together first what he called 'the *physical* – nature, the Cosmos; secondly, the *mental*, including logical and formal abstraction; and thirdly, the *social*'. He was concerned with what he called 'logico-epistomological space, the space of social practice, the space occupied by sensory phenomena, including products of the imagination such

as projects and projections, symbols and utopias' (Lefebvre 1991: 11. The italics are his). Lefebvre's description of his unitary theory aims to integrate the physical world into analyses of the social world. He proposed three levels of analysis. The first he called *spatial practice*, the second the *representation of space* and the third *spatial representations*. In spite of his theoretical insistence on the centrality of the physical, in his methodological approach space and society are equated only with the social, and the physical will not be a dynamic social force, always in flux, that is typical of water. The three levels of analysis inherent in the water-system approach distinguish more clearly the levels that need to be studied, including the physical, the social and the mental. But the way that these three layers or levels are described are all more dynamic, and they are also viewed in conjunction, in order to take the multi-dimensionality of social space and the complexity of historical development into account.

9   Although mainstream social science has been waterblind, in the last few decades a number of studies have been published on water's role in society and history, and good books were also published on the topic early in the twentieth century. The nine-volume series *A History of Water*, presents a kind of state of the art regarding many central issues in the history of water-society relations (see http://www.ibtauris. com/Series/History%20of%20Water.aspx). A number of useful monographies have also been written and here it suffices to mention Worster 1985b, Fagan 2011 and Solomon 2010.

CHAPTER 2: WATER-SOCIETY SYSTEMS AND THE SUCCESS OF THE WEST

1   Several books propose variants of this model of explanation, the most influential being, of course, Weber 1930 [1905].
2   See, for example, Lipsey and Bekar 2004.
3   Shiue and Keller 2004, argue that based on their analysis of three centuries of data from China and Europe, relative levels of market function in China and Europe were similar prior to the Industrial Revolution.
4   For this kind of explanation, see North 1981; North and Thomas 1973; Landes 1998 and Fergusson 2011.
5   Landes 1999: 516. Chapter 3 in the book is entitled: 'European exceptionalism: A different path' (Ibid.: 28–44). This type of explanation has an endless number of forms. Some also explain the difference by 'organisational skills' (see Pacey 1991). Other 'schools' underline different exceptionalist factors. See, for example, the 'Social Change School' (Toynbee 1984 [1884]; Polanyi 1944), the 'Industrial Organization School' (Mantoux 1970 [1928]; Pollard 1964), the 'Macroeconomic School' or the 'Development in Stages School' (see Rostow 1960; Gerschenkron 1962; Fogel 1983) and the 'Technological School' (see Harley 1982). For this assessment, see Crafts 1985. A typical example of how reformers in Britain in the first half of the nineteenth century interpreted the transformation can be seen in the following quotation from 1835:

> In no country have these blessings been enjoyed in so high degree, or for so long a continuance, as in England. Under the reign of just laws, personal liberty and property have been secure; mercantile enterprise has been allowed to reap its reward; capital has accumulated in safety; the workman has 'gone forth to his work and to his labour until the evening,' and, thus protected and favoured, the manufacturing prosperity of the country has struck its roots deep, and spread forth its branches to the ends of the earth. (Baines 1835)

Scholars have explained the revolution in terms of the Enclosure Movement in England during the seventeenth and eighteenth centuries. The exponential increase in food production that followed the enclosure laws that drove the peasantry off the land, increased agricultural production and the urban population undoubtedly stimulated the revolution. Others have underlined the fact that Parliament, unlike the situation with the monarchies of continental Europe, was firmly under the control of the merchant and capitalist classes who were pushing legislation in favour of capitalist growth. Others have pointed to the existence of a new class linked to the Protestant work ethic and the particular status of Dissenter Protestant sects that flourished with the English Revolution. Yet others have focused on particular political-ideological traditions; that the society established after the Civil War in Britain established a link between wealth and status that was conducive to the emergence of capitalist entrepreneurship and development (Perkin 1969). All these factors must be incorporated into any comprehensive study of the Industrial Revolution.

6   See, for example, Williams 1961 [1944], Frank 1969 and Amin 1974 for early advocates of this theory.

7   In one of the most influential contemporary studies (Pomeranz 2000: 107), the author claims that after examining a 'variety of arguments that emphasize internally generated European advantages in productivity before the mid-nineteenth century' he 'found them all dubious'.

8   Pomeranz concludes: 'Far from being unique, then, the most developed parts of western Europe seem to have shared crucial economic features – commercialization, commodification of goods, land, and labor, market-driven growth, and adjustment by households of both fertility and labor allocation to economic trends – with other densely populated areas of Eurasia' (Pomeranz 2000: 107).

9   See also, among many, Flynn and Giráldez 1995: 201–22.

10   For this expression, see Pomeranz 2000.

11   See O'Brien 1987 and Roehl 1976 for the view that France had similar potential for development to Britain in the eighteenth century (see also Crafts 1977 and Broadberry 2007). Commercialisation and specialisation went much further in Britain and the Netherlands than in the Mediterranean states, but Britain began to follow a radically different path from the European river delta with the growth of the British textile industry in the eighteenth century.

12   For an overview of recent discussions between these two schools, see the transcript of the debate between Frank and Landes at Northeastern University: http://www.worldhistorycenter.org/whc/seminar/pastyears/frank-landes/Frank-Landes_01.html.

13   Pomeranz 2000: 34, quoting Adam Smith, *The Wealth of Nations*: 637–8.

14   The seventeenth and eighteenth centuries saw the publication of many books on water control. Fu Zehong, *Xing Shui Jin Jian* (*Golden Mirror of the Flowing Waters*) (1725) and Kang Jitian, *He Qu Ji Wen* (*Notes on Rivers and Canals*) (1804), Jin Fu, *Zhi He Fang Lue* (*Methods of River Control*) (published 1767 [1689]) are all mentioned in Ronan 1995: 230.

15   Mark Elvin, personal communication.

16   Elvin 1973: 286. Elvin also notes examples of making incense and paper, and husking rice in Guangdong, Jiangxi and Fujian during the Qing dynasty.

17   For a description of technology in the iron industry in the eighteenth century, see Dharampal 1971.

18   These issues will be dealt with in a much wider context in my forthcoming book, tentatively entitled: *The Industrial Evolution and the Rise of the Modern World. Why China, India, Japan and the Ottoman Empire failed and some European countries succeeded.*

19   We also need to bear in mind that in Britain, the natural and engineered waterscape eventually proved to be an obstacle to development during the nineteenth century,

as regards both transport and power. Comparative studies of the shift from water power to steam and from waterways to railways could provide new insights into the Industrial Revolution and topics for debate regarding the origins of the modern world.

CHAPTER 3: RIVERS AND EMPIRE

1   For an overview and discussion of the literature on the 'Partition of Africa', see Tvedt 2004a, Tvedt 2004b and Tvedt 2004c. See also the summary of the historiography on the issue in Winks 2001.
2   The interpretation of British Nile policies in the late nineteenth century discussed in this chapter is put in a much broader context in Tvedt 2004a.
3   This expression is taken from Robinson and Gallagher 1961. This chapter quotes from their later edition from 1981. Robinson and Gallagher claim that the overarching British motive in the region was 'Security of the Empire', and that the British became masters of the Nile not because they wanted to, but because they were forced to act by European rivals. What compelled the British to occupy the regions south of Egypt was the fear that other European powers might take control over the Upper Nile as a lever to prise the British out of Suez. The occupation of the Upper Nile was thus seen as a pre-emptive measure by and large forced upon an unwilling and defensive British leadership by other European states interfering in the Nile basin. According to this way of reasoning, the importance of Sudan in British imperial strategy was fundamentally shaped by its conceived role as a buffer state *vis-à-vis* other European powers in the defence of British positions in Egypt, rather than by its intrinsic value to their Nile strategy.
4   Quote from Gallagher and Robinson 1953: 15. More or less identical descriptions of the 'value' of the Upper Nile and Southern Sudan can be found in Sanderson 1965; Holt 1967; Collins 1968a, 1968b, 1969; Brown 1970; Collins 1971; Louis 1976; Sanderson and Sanderson 1981; Collins 1983; Bates 1984; Lewis 1987; Collins 1996; Betts 1972; Pakenham 1992; Cain and Hopkins 2002; Johnson 2003; Webster 2006.
5   This expression is taken from one of the most politically influential hydrologists in the 1890s, William Willcocks. See, for example, his two-volume study of 1899.
6   This was a system of forced labour whereby poor people by the tens of thousands were forced to work to clean and repair the irrigation canals in order to prevent them from silting up, and to help with the maintenance of the canal banks, etc. This system was abolished during the first decades of British rule, mainly due to improvements in the water control system. For a description of the system, see, for example, Willcocks 1899.
7   Willcocks 1894: 5. This bad situation was thus stated in what was a government publication, and newspaper reports from the same year confirm this impression.
8   Quoted in Robinson and Gallagher 1981: 277.
9   Willcocks, *Report on the Nile and Proposed Reservoirs*, in CAIRINT, 3/14/232, NRO. This was written in 1893, and circulated among government officials.
10 Ibid.: 5. The direct gain to the state was said to be from the sale of reclaimed lands and the increase in the annual revenue derived from them. Indirect gain to the state, but direct gain to the country, resulted from the increased value of agricultural produce, the rise in the price of land and in land rents, increase in customs revenue, etc.
11 'Note upon the proposed modifications of the Assuan Dam Project', by W. Garstin, 14.11.1894, Inclosure No. 166, FO 407/126. (FO is Foreign Office and these archival documents can be found in the Public Records Office, London.)

12 'Cromer to the Earl of Kimberley', 15.11.1894, in 'Further correspondence respecting the affairs of Egypt, January to June 1894'. FO/407/126.

13 W. E. Garstin, 'Note on the Public Works Department for the year 1894', 19.2.1895, Inclosure 3 in No. 51, FO/407/131.

14 W. E. Garstin, 'A Note', in Willcocks, *Report*, 1894: 53.

15 See, for example, 'Mr. Rodd to the Earl of Kimberley 3.8.1894', referring to the protest of the London Society of Antiquarians against the proposed Nile reservoir. In 'Further Correspondence respecting the Affairs of Egypt', FO/407/127.

16 Memorandum by Sir William Garstin, Inclosure 1 in No. 30, FO/407/144.

17 W. E. Garstin, 1907: 'Note on the Sudan Irrigation Service', in Inclosure No. 2. *Report of the Finance, Administration and Conditions of the Sudan*, 1906: 53–8. London.

18 W. E. Garstin, 1901: *Report as to Irrigation Projects on the Upper Nile*, in Foreign Office, Blue Book No. 2, 1901 in Despatch from His Majesty's Agent and Consul-General, Cairo.

19 For a description of the role of water-measurement stations in the Sudan for rational water planning in Egypt before 1885, see Chélu 1891: 2–38.

20 See, for example, Milner 1892: 197–8.

21 More than a generation later, the leading Nile expert from about 1920 until the 1940s, H. E. Hurst, summarised what the water planners in the 1890s understood: that the occupation of the Sudan was 'the great landmark' in modern research on the Nile. See Hurst 1927: 440.

22 'For a detailed description of the composition and role of *sadd*, see Rzoska 1976.

23 Lombardini 1865 and Chélu 1891. See also Willcocks 1894: Appendix III, 10–11 and Mason Bey 1881 discussing how removal of the *sadd* could increase the water flow to Egypt.

24 Scott-Moncrieff's own expression in Scott-Moncrieff 1895: 410.

25 R.M. MacGregor: 'The Upper Nile irrigation projects', 3, 10.12.1945, *Allan Private Papers* 589/14/48, Sudan Archives, Durham.

26 In 1890, there were 18 British officials in the Public Works Department, as compared to four in the Financial Department. The Under-Secretary of State, Inspector-General of Irrigation, four Inspectors of Irrigation, three Assistant Inspectors of Irrigation, a Director of Works and eight engineers (List of Appointments held by English Officials, Inclosure in No. 33, Baring to Salisbury, 26.1.1890, FO 407/99.

27 See Chapter 54 on 'Irrigation' in Cromer 1908, II: 456–65.

28 Cromer to Salisbury, 21.10.1891, FFO 141/284.

29 Ibid.

30 Cromer to Salisbury, 14.11.1891, FO 141/283.

31 Cromer to Rosebery, 27.12.1893, Further correspondence respecting the finances of Egypt 1893, FO/407/124. Rosebery answered immediately and supported Cromer's strategy.

32 Cromer to Granville, 3.4.1884, FO 633/6.

33 Cromer to Granville, 21.1.1884, FO 633/6.

34 Cromer to Rosebery, 23.2.1886, FO 633/6.

35 Extract from a minute by General the Viscount Wolseley, Adjutant-General to the Forces concurred in by H.R.H. the Commander in Chief, and forwarded by the Secretary of State for War, 13.1.1890, FO 141/274/16.

36 Cromer to Salisbury, 27.2. 1898, Annual Report for 1898, FO 407/146.

37 'Report by Mr. Garstin on the Province of Dongola', Inclosure in No. 12, Further correspondence respecting the affairs of Egypt, April to June 1897, FO/407/143.

38 Earl of Cromer, 'Report by his Majesty's Agent and Consul-General on the Finances, Administration and Conditions of Egypt and the Sudan', 1899.

39 Cromer's 'Letter of introduction', iii, in Garstin 1904.

40 Earl of Cromer, Report by His Majesty's Agent and Consul-General on the Finances, Administration and Conditions of Egypt and the Sudan 1903, 19.

41 Cromer to Salisbury, 13.3.1890, FO 141/276/84.
42 See 'Correspondence respecting the law-suit brought against the Egyptian government in regard to the appropriation of money from the general reserve fund to the expenses of the Dongola expedition, Egypt'. No. 1 (1897), London: Harrison and Sons, in FO 633/66.
43 Salisbury to Queen Victoria, 25.7.1898, CAB 41/24/42.
44 The influential book by Langer (1936) argues that Prompt in his speech *Soudan Nilotique* made 'some rather indiscreet speculations'. If the water in the Great Lakes reservoirs were not let out in time, the summer supply of Egypt could be 'cut in half'. If the reservoirs were thrown open suddenly and the whole flood sent down to Egypt, the 'civilization of the Nile could be drowned out by one disaster' (Langer 1968 [1936]: 127). He was therefore one actor on the Egyptian scene creating British fear of French intentions, Langer suggests. The entire first chapter of Bates (1984) is devoted to what is described as the threatening visions which Prompt talked about on that fatal afternoon of 20 January 1893. Bates argues further that Prompt had a real influence on French and probably British policy. Collins (1968b) says: 'Prompt did not confine his remarks [in 1893, my comment], however, simply to Nile hydrology. He suggested that a dam constructed on the Upper Nile could destroy Egypt. He who controlled Fashoda controlled Egypt' (Collins 1968: 16). Collins also writes that Fashoda 'had long been considered the hydrological key to the basin of the Upper Nile', and 'the point where the Nile waters could best be controlled' (Ibid.: 4). This story of the fears that Prompt created is repeated by Collins (1996), and the story of Victor Prompt's role is also mentioned in general books on the theory of peace and war (Brown 1998: 200).
45 One was entitled 'La Vallée du Nil', and was given on 6 February 1891; another lecture was 'Note sur les réservoirs d'eau dans la Haute Egypte', on 26 December 1891; then came the notorious 'Soudan Nilotique' on 20 January 1893, and finally 'Puissance électrique des cataracts' on 28 December 1894.
46 Ibid.: 44. The text reads: 'Le doute n'est plus permis, et il faut reconnaître que l'Égypte d'aujourd'hui est menacée dans toutes ses richesses et dans son existence même, par la nature des choses, sans avoir besoin, pour cela, de supposer que les riverains au-dessus de Wady-Halfa peuvent utiliser l'eau d'étiage et en priver l'Égypte absolument'.
47 Ibid.: 48.
48 Ibid.: 51.
49 Ibid.: 56–8.
50 Ibid.: 60.
51 Ibid.
52 White 1899. Silva White communicated with Wingate and was familiar with British policies.
53 See, for example, Willcocks 1894, Appendix III. He wrote '[...] all the small ponds and pools cease to aid the stream, and if they are very extensive, as they are south of Fashoda, they diminish the discharge considerably by their large evaporating areas', and he dismisses Prompt, described as a railwayman, and his findings, as the findings of a layman (Willcocks 1894: 17). These speculations were left out of the official report published in the following year, but while Willcocks' ideas were rational and realistic, Prompt's were irrational and not very realistic. Magnus (1958: 138) also repeats the idea that Fashoda 'was regarded hydrographically as the key-point on the Upper Nile'.
54 Brown simply misunderstood the nature of the geography and hydrology of the Nile: 'The strategic centre of this region was the ancient fort of Fashoda at the headwaters of the Nile.' Fashoda was of course no such strategic centre. See Brown 1970: 23.

55 See Tvedt 2004 for a more detailed discussion of the importance of European rivalry in the valley.

CHAPTER 4: RELIGION AND THE ENIGMA OF WATER

1 The extensive literature on the Storm-God includes deities that also might be seen and interpreted as 'water gods' or rain gods, since the wind would tell the people that clouds were on their way and that these clouds brought what was most crucial of all: rain. Rain was important in all agricultural civilisations, of course, but it was more important in dry areas. A discussion of how such gods are classified might reveal interesting new relationships between god-making and environment (see Green 2003).

2 In Weber's book there is not a single word about the role of water or rivers in religion, although he is discussing Christianity, Hinduism, Judaism and Islam.

3 See Shaw 2006 on the possible role of archaeological studies of landscapes and water in India.

4 Tvedt and Oestigaard 2006 and Tvedt and Oestigaard 2010, discussed a wide range of research on the role of water in religions and religious rituals (see, for example, Khasandi-Telewa 2006; Armstrong and Armstrong 2006; Kodiyanplakkal 2006; McKittrick 2006; Namafe 2006; de Châtel 2006; Oestigaard 2006; Gerten 2010; Strang 2010; Sætersdal 2010; Zheng Xaio Yun 2010; Doniger 2010).

5 Mircea Eliade wrote: 'Water symbolises the whole of potentiality: it is the *fons et origo*, the source of all possible existence [...] water symbolises the primal substance from which all forms came and to which they will return' (Eliade 1979: 188). The point of this article is not limited to demonstrate the importance of water in religions, but to suggest approaches to the comparative study of religion.

6 Mimir's well was the source of wisdom in Norse religion, and Odin, the principal god, was willing to sacrifice his eye to the waters of the well in exchange for its wisdom. The act in the well was a kind of 'axis mundis' for Odin's power and the maintenenance of cosmos. Water in the Viking cosmology was conceived of very differently from the water in the desert religions, and was more similar to the traditions of the Celts. There is disagreement regarding how to interpret the fact that people in pre-Christian Norway were also sprinkled with water when they were named and became part of the clan. Some argue that it reflects influence from the religions that dominated Southern Europe at the time, while others regard it as a kind of universal ritual that also happened to be practised by the Vikings. A hypothesis could be that it should not be seen as a universal ritual although it spread to many areas of the world, and that it came to Northern Europe after year 0.

7 The importance of rainmaking cosmology in the Norse religion is yet to be studied properly and within a comparative perspective. In Snorre's sagas of the kings it is, for example, described how King Domalde was sacrificed by his chiefs after three bad years in a row.

8 An exception is Best 1999 who, after assessing documents and records about the floods in the region concluded that it is not history, but a mythologised legend, and that it is based on a local flood where a trader called Noah lost his boat.

9 In Narayan and Kumar 2003 ecology is defined as '(a) physical environment which includes the non-living entities consisting of chemical and biological components, and (b) biological environments consisting of living organism (plants and animals) as called biosphere. The oceans, lakes, streams, rivers form hydrosphere and the air envelope that surrounds the earth is called atmosphere' (xix).

10 See also Hargrove 1986; Holm and Bowker 1994; Gottlieb 2003. See also the literature list for the case studies published by the Harvard programme on religion and ecology.
11 See also the discussion in Ayres 2001: 346–53 of the role in the second century AD of Novatian, whose followers were called Cathars or Puritans. He deferred his baptism because he was afraid of the sins he would commit after being baptised. Baptism washed away all sins, but sins after baptism were considered almost unpardonable. Eventually he became sick and, being near death, had to be baptised lying in his bed.
12 Personal observation, also described in Tvedt 2014a.

CHAPTER 5: BETWEEN THE HYDROLOGICAL AND HYDROSOCIAL CYCLE

1 Herodotus, 1960.
2 As the ancients were well aware, water was not the only essential. Without air, death takes place in about five minutes. Without water, it takes three to four days. But the important thing in a historical and social perspective is that air cannot be tamed, piped, controlled or diverted (although in practice this is done for a number of specific purposes such as scuba diving, submarines, firefighting, space travel, etc.). But in a city there is never too much of it, and it is not changing forms and in a constant flux in the same way as water. So the parallel to water, from the point of view of urban development, is not very interesting, except when it carries 'alien' matter and becomes a major pollution problem as it did in "smoggy" London' in the early twentieth century and in Beijing in the early twenty-first century.
3 Gradually more and more empirical studies have been carried out on the relationship between water and urbanisation, but the focus and perspective of these studies have not been integrated theoretically and conceptually into urban studies. Blake 1956 is a seminal pioneering work on one aspect of the water/city nexus in the USA. Another very interesting and important work is Guillerme 1988 [1983]. It analyses the history of hydraulic systems in French cities of the Middles Ages. He argues that the hydrographic system did not originate naturally but that its gradient, profile and dimensions were planned and worked out over many centuries – from as early as the decline of the Roman Empire – for military purposes. In time, this system gradually fulfilled a number of other and variable tasks.
4 Hall, in his summary of the urban geography tradition, argued along the same lines as Carter had already done when criticising site-and-situation geography. While Carter wrote that this 'site-situation approach was meaningless when the large urban conglomerates had to be considered' (Carter 1981: 4), Hall wrote that the development of cities itself had made this tradition obsolete: 'Original location factors have tended to be overridden by the scale of subsequent urbanisation or have greatly declined in importance as the form and function of urban areas have changed' (Hall 1998: 20).
5 Urban morphology, which for some years was very popular and according to Herbert and Thomas perhaps the only truly 'indigenous' line of evolution within human geography (see Herbert and Thomas 1982: 12) has also displayed minimal interest in water/urban links. Urban morphology – the study of the city as a human habitat – initiated many interesting research projects and publications on urban forms, but the whole concept of urban form was considered as being firmly rooted in social and economic processes only (see Whitehand 1977). Urban morphologists focus on urban forms perceived as the tangible results of social and economic forces, and the confluence of city and water is beyond its scope.

6 'The functioning of the city as a system, the concepts of growth and decay for example, can be related to processes found in a natural ecosystem. An investigation of cities based on analogy to the physical ecosystem is a procedure that relates all parts to the whole' (see Exline and Larkin 1982: xvii).

7 A bibliography of urban geography research between 1950 and 1970 as reflected in the articles and book reviews of 72 US, Canadian, British, Dutch and Scandinavian geographical periodicals listed 2,949 publications. Of them 10 dealt primarily with water issues and a few studies also dealt with the role of rivers for the location of cities and towns (see Strand 1973).

8 Different historians criticised Worster's delineation of the topic. Martin Melosi criticised these views in *Environmental History Review* in 1993, arguing that it was illogical for environmental historians to study farming and not the history of cities and towns (Melosi 1993: 4.) and Samuel Hay wrote that the environmental historian should study how people had acted upon the environment over time wherever it was, including in relation to urban history and development. For Joel A. Tarr urban environmental history was primarily the story of how man-built or anthropogenic structures ('built environment') and technologies shape and alter the natural environment of the urban site with consequent feedback to the city itself and its populations (Rosen and Tarr 1994). Melosi suggested a broader definition in which the physical features and resources of urban sites (and regions) influence and are shaped by natural forces, growth, spatial change and development, and human action. Thus the field should combine 'the study of the natural history of the city with the history of city building and their possible intersections' (Melosi 1993: 2).

9 Others have suggested a theoretical and methodological justification for doing urban history in a broader sense, including analysing the interplay between city and the physical environment. Urban history should be 'the study of the role and place of nature in human life', and 'illuminate the dialectic interdependence between cities and nature' (Rosen and Tarr 1994: 307). This call helped to open up the field and it coincided with the popular environmentalist idea about nature as a unifying force. But for individual research projects it is an impossible task. No historian will ever be able to accomplish such a study because both culture and nature are two terms that cover too much. The articles that were presented in their volume on *Urban History* did not discuss all these relationships, either, but mostly, in fact, how cities impacted on the local waterscapes.

10 In 1980, R. J. Johnston wrote his outline for urban geography. It should be concerned with the 'organization of space within an urban context'. This space was, however, defined as socially constructed. 'The urban system,' he wrote, reflected and interacted 'with the 'economic, social and political structure of the society which occupies it', while nature and water was left out of the picture (Johnson 1980: 26). A couple of years later, another book on urban geography summarised the way ahead: 'The urban geography of the future will surely be more firmly embedded in the wider philosophies of social sciences than has hitherto been the case [...]' (Herbert and Thomas 1982: 25). Influenced as it was by post-structuralism and post-modernism, it was now argued that the term nature itself should be discarded, since nature as such was a social or cultural construct. In urban geography this trend encouraged the 'sociospatial' perspective, rejecting the relevance of physical factors, being instead solely concerned with how human experience gave meaning to and created space (Gottdiener 1994; Harvey 1996a and 1996b).

12 See, for example, Doyle 2009.

13 See, for example, Diefendorf and Dorsey 2005; Plat 2005; Weiner 2005; Davis 2006; Swyngedouw 2004.

14 Lees and Lees 2007.

15 See, for example, Kaika 2003 and 2005; Swyngedouw 1996, 1997, 1999 and 2006.

16 Roman water engineering for public supply and hydro-therapeutic use set the standard by which urban water systems were to be judged well into the early modern period. Medieval systems for controlling and channelling water were less grandiose, and did not involve complex technologies to transport water over long distances. Instead, medieval water control in Europe tended to be local and open, often serving as defensive as well as a supply system in the case of moated towns. As Guillerme notes, the French medieval city 'knew how to master the hydraulic environment, and it is precisely on this point that these cities differed fundamentally from the earlier Gallo-Roman or the later industrialised city, both of which dreaded surface water (Guillerme 1988 [1983]). Monastic orders were leading the way in water control technology and aquaculture in early medieval Europe, and examples have been found of water supplies being extended from monasteries into local towns.

17 See Tvedt and Oestigaard 2006 and 2010 for two volumes containing many case studies aimed at encouraging this type of research.

CHAPTER 6: WATER, SOVEREIGNTY AND THE MYTH OF WESTPHALIA

1 A legal definition of sovereignty in *Black's Law Dictionary* reads that sovereignty is the supreme domination, authority or rule and also the supreme political authority of an independent state. Sovereignty has an external and an internal aspect – the external being the power of dealing on a nation's behalf with other national governments, the internal one being the power that the ruler exercises over his subjects. A sovereign state is the state which possesses an independent existence, which is complete in itself without being merely part of the larger whole to whose government it is subject. It is a political community whose members are bound together by the tie of common subjection to some central authority, whose commands those members must obey (1999: 1402, 1401).

2 Croxton and Tisher 2002 summarised the problem of historical analyses of Westphalia very accurately: 'Taken together, the congress and the peace are so complex that historians still discover new aspects of it today' (xx). This article highlights only one aspect of this whole process.

3 'Treaty of Westphalia. Peace Treaty between the Holy Roman Emperor and the King of France and their respective Allies', see http://avalon.law.yale.edu/17th_century/westphal.asp. The Yale translation talking about the 'advantage of 'the other' is less accurate than the old English translation, but what is significant is that the principle that the sovereign should work for the advantage of each other is common to both of them.

4 'The Articles of the Treaty of Westphalia. Peace Treaty signed and sealed at Munster in Westphalia the 24th October, 1648'. 1697. London: W. Onley.

5 'The Articles of the Treaty of Westphalia. Peace Treaty signed and sealed at Munster in Westphalia the 24th October, 1648'. 1697. London: W. Onley, 31.

6 I was made aware of some of these points by Pierre Beaudry in connection with work on Vol. III, Series II of *A History of Water*.

7 For a description of these difficulties, see Mellor 1983.

8 Cioc 2002: 32.

9 Mellor 1983: 70.

10 On the other hand, the agreement closed the River Scheldt to the Belgic provinces, thus apparently ruining the commerce of Antwerp (Kaeckenbeeck 1920: 31), an

expression of the fact that the agreement was less concerned with principles than with pragmatic solutions suiting the most powerful.

11 First after the next big European peace conference, the Vienna Congress in 1815, and after the French Revolution had swept away the old order, partly by establishing the Confederation of the Rhine in 1806, did the countries in the region succeed in developing the Rhine as a transport artery. Then it took place under the leadership of the famous water engineer Johan Gottfried Tulla, who increased the depth of and channelled the Upper Rhine. This remodelling of the Rhine also required, of course, a technological level in river manipulation that was not available in the seventeenth century.

12 Similar conclusions have recently been drawn by historians researching other aspects of Westphalia. The Thirty Years' War was accompanied by permanent negotiations and the opponents never totally broke off political contact, and ideas of mutual destruction did not exercise a decisive influence over the political elites (Kampmann 2010: 204).

13 The point about borders and their increasingly symbolic functions is made by Rudolph 2005.

14 Dinar 2008 is one of the few books underlining the importance of geography for understanding water law, but the approach and explanations are very different from the suggestion put forth here.

15 The role of the Danube Commision was so strong that an observer in the 1930s argued that 'the need for protecting the integrity of the commission will some day lift it out of the twilight of statehood and accord it full membership in the League of Nations'. See Blackburn 1930.

16 Quoted from Bourne and Wouters 1997: 4.

17 Quoted from Bourne and Wouters 1997: 4.

18 Quoted from Bourne and Wouters 1997: 15.

CHAPTER 7: WATER AND INTERNATIONAL LAW

1 The Nile Waters Agreement was for a long time hailed as a model. The following quotation is typical: '[...] the Anglo-Egyptian negotiations over the Nile afford an admirable example of how such problems should be envisaged and solved. Here we have no dogmatic insistence upon abstract doctrines of territorial sovereignty or riparian rights, no claim to disregard the history of an ancient land or to veto the proper development of a new' (Smith 1931: 70).

2 Lord Lloyd to Mohammed Mahmoud Pasha, 7.5.1929, in *Sudan Pamphlets* 89.

3 See especially Garstin 1899, 1901, 1904 and Dupuis 1904. See also Cromer 1908.

4 The Blue Nile's annual innundation in Khartoum is on average estimated as 54bcm/yr; at Aswan it is 48bcm/yr. Ethiopia's contribution to the Nile system through the three head streams – i.e., the Blue Nile, Tekeze-Atbara and Baro-Akobo river basins – is 68.7bcm/yr or 82 per cent of the total Nile flow.

5 Foreign Office Memorandum, Sperling, 'Resumption of negotiations for the construction of a dam on Lake Tsana', 8.11.1922, FO 371/7151.

6 Foreign Office Memorandum, Murray, 4.1.1923, 'Memorandum on the political situation in Egypt', FO 371/8972.

7 Kewin-Boyd to Allenby, 14.3.1920, FO 371/4984.

8 Allenby to Austin Chamberlain, 15.12.1924, FO 371/10046.

9 Quoted in Kurita 1989: 26.

10 Quoted in Sir John Maffey to Sir P. Loraine, 15.8.1930, FO 371/4650.

11 The crisis in Egypt. Mr MacDonald on the ultimatum. A mandate for the Sudan, *The Times*, 29.11.1924.

12 Ziwer Pasha to Allenby, 26.1.1925, 'Texts of notes exchanged between Lord Allenby and the Egyptian Government on January 26th, 1925, regarding the control of the Nile water'. FO 371/10882.

13 Allenby to Ziwer Pasha, 26.1.1925 (as put together in the Foreign Office from Cairo telegrams), FO 371/10882.

14 Note from Lord Allenby to Ziver Pasha, 26.1.1925, in *Sudan Pamphlets* 89.

15 MacGregor found the Egyptian member difficult to co-operate with. In order to come up with an acceptable report, he informed Allenby and the British Government that he had had what he himself called clandestine meetings with the British water-planners Hurst and Butcher, who were employed by the Egyptian Government (Allenby to Chamberlain 25.5.1925, Enclosure 3 in No. 1 by Mr. R.M. MacGregor).

16 Nile Commission 1925: 30.

17 Ibid.: 28.

18 MacDonald's *Nile Control* provided, during the period when the Sennar reservoir would be in use, a water allowance at canal head of 15 m³ per feddan per day, including 33 m³ for losses between canal head and the 5,000 feddan blocks. MacGregor had worked out, on the basis of figures obtained from research at Hag Abdulla and Wad-el-Nau, that a water allowance at canal head of 10 m³, including two cubic metres for losses, would suffice. Thus, only two-thirds of the water provided would be actually required, and an extension of 150,000 feddans became possible on the assumption that the reservoir drew upon the Nile from 18 January to 15 April. In terms of volume, this saving amounted to 5 cubic metres per feddan per day on 300,000 feddans for 87 days, i.e., 130.5 million m³. Moreover, *Nile Control* argued that the date from which the canal would have to be supplied from storage was 18 January and the waters could be brought back to the river at the end of March. MacGregor discovered, however, that the former date should be brought back to nearer the beginning of January, which also made it possible to bring the latter date forward to the beginning of March. Therefore it was assumed that the reservoir would be called upon to serve the present area for a period of 60 days instead of 87 as contemplated in *Nile Control*. This saving would amount to 15m³ per feddan per day on 300,000 feddans for 27 days, i.e., 125 million m³. Assuming the period to be 65 days, the volume available would permit an extension of 190,000 feddans.

19 Lord Lloyd to Sarwat Pasha, 16.2.1928, Enclosure 1 in No. 1, Lloyd to Chamberlain, 23.2.1928, FO 371/13138.

20 Draft of a Note to be addressed by His Majesty's High Commissioner to the President of the Council of Ministers, Enclosure 3 in No. 1, Lloyd to Chamberlain, 23.2.1928, FO 371/13138.

21 He wrote a note about the Egyptian Government's consideration of the report of the Parliamentary Finance Commission on the budget of the Irrigation Department for the current financial year. Under the heading 'Sudan', an estimate of £E1,100,000, of which £E130,000 was to be spent in 1928, had been included for the 'modification and improvement of the flow of the Nile in the sudd region', by means of large dredgers to be purchased abroad. Lloyd to Chamberlain, 12.5.1928, FO 13138.

22 Foreign Office minute, Murray, 1.8.1928, FO 371/13138.

23 Foreign Office to Lloyd, draft, 15.3.1928, FO 371/13138.

24 Draft Note, Lloyd to the Egyptian Minister for Foreign Affairs, n.d. [July 1928, my comment], FO 371/13138.

25 Lloyd to Chamberlain, 14.7.1928, FO 317/13138.

26 Lloyd to Chamberlain, 20.2.1928, FO 317/13137. The Nile Board that should be responsible for the entire Nile should be made up of two representatives of the Egyptian Government and two representatives chosen by His Majesty's Government (Allenby and the Governor-General of the Sudan, Maffey, agreed that they should represent the Sudan Government, and the salaries should be paid by Khartoum).

27 Lloyd to Chamberlain, 14.4.1928, FO 371/13138.
28 Copy of letter dated 12 April 1928 from the Secretary of the Aboukir Company, Ltd., to his Excellency the Minister of Public Works, FO 371/13138.
29 Draft letter, Foreign Office to Sir W. F. Gowers, November 1929, FO 371/13857.
30 Parkinson, Colonial Office to the Under-Secretary of State, Foreign Office, 2.11.1929, FO 371/13857.
31 They had just organised fisheries surveys in these lakes for the first time (see Worthington 1929).
32 Parkinson, Colonial Office to C. J. Norton, Foreign Office, 14.11.1929, FO 371/13857.
33 Murray to the Under-Secretary of State, Colonial Office, 2.12.1929 FO 371/13857.

CHAPTER 8: WATER-SOCIETY RELATIONS AND HISTORY OF THE LONG TERM

1 The term waterzone might be useful as part of such a research strategy, especially when it comes to agricultural development. Waterzone is a broad water-geographic division, based on distributional differences of precipitation and run-off characteristics that have a fundamental importance on biodiversity, flora and fauna and hence on potential for rural organisation, and covering small or large areas depending on characteristics of the waterscape of differential importance for societal development. The idea is based on the assumption that the probability of encountering similar types of agricultural practices in any given waterzone remains relatively constant, and within an acceptable range of variation. The concrete deliniations must be adapted to what is fruitful in order to understand local circumstances. It is a more complex term than an ecozone because the latter is only concerned with 'the natural' while the waterzone must also reflect man-made changes to the waterscape. But it is much more empirically managable than an 'ecozone' which covers basically everything. Each waterzone shares a large majority of both species and agro-rural dynamics, and since the confluences between water and society are quite similar, the interactions also show patterns of similarity that might be reconstructed and analysed.
2 It is not mentioned in Nordström 1986 or in Imsen and Winge 1999. In these books dealing with the dictionary of Scandinavian history there are no entries on either 'drenering' [drainage] or 'grøfting' [trenching], while there are entries on relatively esoteric topics such as 'drengebåt' [boat belonging to a merchant in Bergen], 'drette' [to raise cattle/ drag a harrow], 'drinke [tynt øl]' [to drink very light beer], while instead of 'grøfting' there are entries like 'grønnsaltet' [extremely heavily salted meat] and 'grøtstein' [soapstone or steatite]. In fact, there is no entry on water at all, but room has been found for entries on 'vannfarken' (water-gypsies). The issue is not once discussed in the most influential multi-volume series on Norwegian history edited by Knut Mykland. Kåre Lunden in his volume on the history of Norwegian agriculture explicitly stated that draiange was of no importance in Norwegian agricultural history (Lunden 2002).
3 The only places in Norway where there has been some form of water shortage are in the Lesja and Sjåk areas in the upper parts of Gudbrandsdalen. Here the annual precipitation could be less than in parts of the Sahara. The area never became a desert, of course, because of the melting of snow in the mountains and because of the modest evaporation and the ability of the soil to retain the water. These are the only villages in Norway that for hundreds of years depended upon irrigation.
4 Jan Lindegren (1988) described the interest that King Gustav Vasa took in ditch digging for Sweden. He warned that the whole country was going to ruin and would become bog and marsh, where 'wild rosemary, bog myrtle, bilberry and cranberry' grew, all because 'the fields have not been held by ditches as they should'. The exhortation was clear: 'drain your fields'.

5 The Egyptian Prime Minister in 1890 summarised Egypt's predicament: 'The Egyptian question is the irrigation question' (quoted in Tvedt 2004),

6 See Bloch 1935; Forbes 1956; White 1978; Gimpel 1988; Reynolds 1983.

7 Dyrvik et. al. 1979 describes Norwegian economic history 1500–1850 and also discusses Norwegian agriculture. It fails to include or integrate the water-society nexus in the perspective. The implication is that the book fails to explain fundamental and particular preconditions for early Norwegian mining industries, timber export, fishing and agricultural development.

8 For some sources for this chapter's analysis of Norway, see Tvedt 1997.

9 Willcocks 1889.

10 Williamson and Panza 2013. Landes 1991: 59 describes Ali's industrialisation effort as 'a project that was doomed from the start and already in its death rattle'. The reasons Landes gave were only political and cultural. Batou describes him as a '*roi industriel*' (Batou 1993: 94), but without explaining why he failed by other than British opposition.

11 In spite of the fact that the British at the beginning of the twentieth century could not dam the entire flow of the Nile – mainly due to the amount of silt in the Blue Nile flood – they could have harnessed the river for power production on a smaller scale. But in addition to the hydrological challenges of the Nile, river control presented a water-political dilemma: the needs of agricultural irrigation required uneven flow and more water in the summer season, while power production required a quite even flow all year.

12 This was the expression used by Lord Cromer, the British ruler of Egypt between 1883 and 1907.

13 Holmsen 1961: 261.

14 See Lunden 1981.

15 See Lipset and Rokkan 1967.

16 For this discussion see, for example, Wittfogel 1981 [1957] and Butzer 1976. Wittfogel argued that a distinctive type of political system, absolutist and bureaucratic in nature, tended to develop in arid or semi-arid regions which made the transition from hunting and gathering to agriculture. Butzer refuted this thesis, arguing that it had been falsified by empirical findings, especially in Egypt.

CHAPTER 9: WATER AND CLIMATE CHANGE

1 For some relevant literature, see World Water Assessment Programme 2003; Kabat et al. 2003; Nijssen et al. 2001; Arnell 1999; Frederick and Major 1997; Major and Frederick 1997; Boorman and Sefton 1997; Rind and Goldberg 1992; Loáiciga 2003; Stefan et al. 1998; Qin and Huang 1998; Bonell 1998; Kabat et al. 2013.

2 Scientists from the University of Potsdam, Germany and the GFZ German Research Centre for Geosciences analysed organic remains extracted from Meerfelder Maar sediments from the Eifel region in western Germany, to reconstruct changes in precipitation patterns in unprecedented detail.

3 Von Köppen also discussed the role of temperature, but a close look at his classification system makes it clear that it is water in all its different forms that was the most fundamental basis for his system.

4 It deals with the soil water budget using evapotranspiration, monitoring the portion of total precipitation used to nourish vegetation over a certain area (employing humidity and aridity indices to determine an area's moisture regime). Thornthwaite co-authored with Mather the monograph 'The water balance' in 1955.

5 WHO 2013 presents data on new research findings about climate change, but offers no direction for further research. It represents the perspective and projections and

scenarios of WHO, but without explaining or justifying them. Given their focus and institutional interests, it should be hardly surprising that the report does not mention the role of water in climate.

6 It is argued that another 'uncertainty' will be more important for the future of the water issue, at least in the short run. Impending global-scale changes in population and economic development will dictate the relationship between water supply and demand more than will changes in mean climate (Vörösmarty et al. 2000).

7 Two examples of how different opinions of the water landscape of tomorrow are turned into contemporary politics: When the conservative Italian Government under Prime Minister Silvio Berlusconi decided to build the Mose Project to save the lagoon of Venice, it did so based on opinions about the predictions of how the water level would rise in the future due to global warming. The Green Movement in Venice regarded the building of the barriers as an ecological catastrophe, partly on the basis of alternative assessments of climatic trends. The Greens questioned the idea that the sea level would necessarily rise. In India, governments of different states have objected to the National Water Link plan that aims to connect many of the large river systems into one national water 'grid'. They claim that they may well have surplus water at the present, but due to the uncertainties regarding tomorrow's water landscapes they have no excess water to give away permanently to water-deficient states.

8 In the Framework Convention on Climate Change, climate change refers to 'a change of climate that is attributed directly or indirectly to human activity that alters the composition of the global atmosphere and that is in addition to natural climate variability observed over comparable time periods'.

9 Wheeler and von Braun 2013.

10 There are surprisingly few studies of the importance of snow, in spite of its importance for economic patterns and agricultural cycles, and the fact that snow is one of the true wonders of nature, unique in the forms in which its crystals fall from the skies and remarkable in the way it changes not only the landscapes but also urban centres in a matter of seconds.

11 The IPCC special report on climate change adaptation estimates that around 1 billion people in dry regions could face growing water scarcity in the decades to come.

12 Capon et al. 2013.

13 Taken from Oliver 1973: 27.

14 Hassan 1997.

CHAPTER 10: A CRITIQUE OF THE SOCIAL SCIENCE TRADITION

1 Weber and Durkheim were editors of the two major sociological journals at that time: *Archiv für Sozialwissenschaft und Sozialpolitik* and *L'Année* respectively. Both journals derived their basic approaches from intellectual traditions as different as French rationalism and German historical thinking, although they still agreed on the issue of the social. In his book on Durkheim's influence, Giddens omits to discuss his most influential idea – that of delineating the object and task of social sciences to that of social factors (see Giddens 1986). Giddens in his *Human Society* ends his introduction thus: 'Sociological thought must take an imaginative leap beyond the familiar, and the sociologist must be prepared to look behind the routine activities in which much of our mundane life is enmeshed' (Giddens 1992: 4). When it comes to societies' relation to nature, Giddens as a theorist was conventional; in this work on human societies in general there was no chapter or index entries on these topics.

2   See Durkheim 1893 (1984): 285–6 for this assessment.

3   There is general agreement about this assessment. Marx's ideas here build on Hegel who held that freedom, or humanity's realisation that it is free, entails its recognition that it is separate from and sovereign over nature, where 'nature' is both the external environment and the natural inclinations of man himself.

4   Especially in his later writings one can find formulations suggesting a productivist logic that sees nature as an enemy opposed to man, as a resource and object that must be mastered, exploited and controlled. Marxian ecologists argue, however, that there are ideas of an alternative society–nature relationship and of nature conservation in his writing.

5   In footnote 6 of *Das Kapital*, Vol. I, Marx quotes: 'The necessity for predicting the rise and fall of the Nile created Egyptian astronomy, and with it the dominion of the priests, as directors of agriculture' (Cuvier 1964: 141). Here, in one sentence, Marx argues that the whole social fabric of classical Egypt can be explained not by social relations or productive forces, but by the irregularities and regularities of a river's fluctuations or hydrology. This way of integrating physical waterscapes in the analysis is not followed up or systematised, and it was not made an object of theoretical reflection. See Marx 1906 [1867].

6   The debate about this Marxian concept has lasted for many decades. A common trait of most discussants is that they have only raised issues of a sociological nature. We can take Krader as an example. His point is that the nature of the state and the nature of unfree labour have to be understood (Krader and Kovalevskii 1975; Fogel 1988, who debates the situation in Soviet Russia, China and Japan).

7   Weber 1978. See the paragraph on 'Types of Technical Division of Labour', where Weber lists these divisions and mentions watermills but never discusses how different waterscapes established different potentials for this form of differentiation of labour (Weber 1978: 120–1).

8   Hippocrates wrote over two millennia ago in *Airs, Waters, Places:* 'For in general you will find assimilated to the nature of the land both the physique and the characteristics of the inhabitants' (Hippocrates 1923: 137), and scholars from Aristotle, via Montesquieu, to Hegel had attributed cultural and psychological differences to dissimilarities in environment. Sociologists like Herbert Spencer and zoologists like Huxley based their works mainly on analogical methodologies that would make it possible to investigate the characteristics of nature and of societies and human beings contemporaneously, by assuming that these phenomena were regulated by the same laws of behaviour and of evolution.

9   See also Edward Soja. He summarised that the emergence of the social sciences was closely tied to the disappearance of physical space as a category of knowledge accumulation: 'the explicit theoretical rejection of environmental causality and all physical or external explanations of social processes and the formation of human consciousness. Society and history were being separated from nature [...] a relative autonomy of the social from the spatial' (Soja 1989: 34–5).

10 Durkheim exerted an 'extraordinary influence over the development of modern social thought' (Giddens 1986: 7). His conception of social science became infused into a number of disciplines.

11 The author went through all the relevant books in the University Library, Cambridge, in April and May 2013.

12 Eric Hobsbawm's famous book about the twentieth century is entitled *The Age of Extremes: The Short Twentieth Century, 1914–1991*. It was published in 1994.

13 Donna Haraway also suggested dropping the nature–society dichotomy in favour of a sociology of *hybrids* (see Haraway 1991).

14 Two of the most influential thinkers about society after World War II, Habermas and Luhmann, shared this disregard for nature, but in different ways. On the one hand,

Luhmann argued that only when nature became an object of communication did it exist for society and then as a communicative product. On the other hand, Habernas in his attempt to reconstruct his version of Marxist-inspired historic materialism tried to replace the concept of labour, which by Marx was tied to nature, with the concept of interaction, which was tied to morality. Historical development was now seen as a result of a moral evolution and as an outcome of social self-organisation through interaction (Habermas 1979).

15 Swyngedouw 1999.

16 Beck writes: 'Nature is not nature, but rather a concept, norm, memory, utopia, counter-image. Today more than ever, now that it no longer exists, nature is being rediscovered, pampered' (Beck 1995: 65).

17 Huntington assembled an imposing mass of data and did much to establish the fact that there had been significant post-glacial changes to climate, and he was one of the first to write about the ozone layer and the passage of cyclonic disturbances, but he also wrote 'The character of races', in which he discussed from a racial perspective the origin of 'civilized man'.

18 Huntington wrote: 'England, as we have seen, probably has the best climate in the world. It keeps people out doors, and makes them tough and sturdy; it stimulates the mind, and makes it easy to think clearly and act energetically. Thus when the British are pitted against other nations their extra energy has again and again turned the scales and enabled them to hold parts of the world against their rivals' (Huntington 1922: 395). Many versions of this idea have been put forward: 'No one believes it to be an accident that the progressive and energetic peoples who dominate the world are found in the intermediate zones' (Whitbeck 1932: 87).

19 He argues:

> The variations in the rate of human progress from region to region and century to century, however, can be understood more readily. Just as the windings of the motorist's road are influenced by mountains and rivers, so the march of civilization is influenced by man's physical environment as a whole. On the other hand, the sharpness of the curves, the smoothness of the tire roadbed, and the steepness of the grades are cultural qualities. They depend on man's work and correspond to the cultural background against which historic events take place. Finally, the element of biological endowment is represented by the character of the motorist, on the one hand, and of the people in any historic epoch, on the other. Thus, the course of the car depends on (1) the general direction of the road; (2) its curves and grades; (3) the quality of the road; and (4) the quality of the driver – that is, upon all four of our factors.

20 The re-emergence of evolutionary theory in archaeology came about with the neo-evolutionary theories of V. Gordon Childe, Julian H. Steward and Leslie White (Eddy 1991). The opposition to cultural stereotypes and racism stimulated the interest in ecology. Through a synthesis of the original concepts of classical cultural evolutionism, the Historical school, and the British social anthropologists' Functionalist approach (as represented by the theories of A. R. Radcliffe-Brown and B. Malinowski), neo-evolutionary theory was developed (Eddy 1991; Wolf 1964; Garbarino 1977).

21 Axelrod 1984; Bartlett 1990; Beach et al. 2000; Born and Sonzogni 1995; Nunes Correia and da Silva 1999; Dufournaud and Harrington 1990; Keiter 1994; Falkenmark and Lundqvist 1999; Frey 1993; Hardin 1968; Martin 1994; Priscoli 1990; Starr 1991; Homer-Dixon 1999; Morrissette and Borer 2004; Arsano 2007.

22 Hardin 1968: 1244.

23 For the many works on Nile discharges, rainfall and other hydrological phenomena by H. E. Hurst and his colleagues in the Egyptian Ministry of Public Works, and for

other works on these subjects in general, see Tvedt 2004c, which lists almost 1,000 entries on the physical characteristics of the river.

24 The extensive literature assessing Hardin's thesis includes Aggarwal and Narayan 2004 and Ostrom et al. 2002.

25 For Tanzania and Kenya, see Ngowi 2010 and Mwiandi 2010.

26 See Tvedt 2014 for a more detailed discussion of such differences along the River Nile.

27 Rivers have been understood in the same way as John Locke's famous discussion of how a tree continues to be the same thing, even after its parts have changed. A tree differs from its mass of matter, since the mass of matter changes and its parts are constantly shed and replaced. Yet the tree is still the same tree; it persists despite its changing composition, because its mass is constantly organised to maintain the life of the tree (Locke 1910: 240–1).

28 Braudel used this analytical framework in Braudel 1975. He also made the point that the three levels are a conscious simplification: 'History exists at different levels, I would even go so far as to say three levels but that would be only in a manner of speaking, and simplifying things too much. There are ten, a hundred levels to be examined, ten, a hundred time spans' (Braudel 1980: 74).

29 Another issue of time and water, but related to our discussion, is how changes in a society's management of water can 'create' new social time. Since water is a dire necessity for all people every day, how water is provided and how it is brought to the 'individual' consumer are two of the factors with the most bearing on the time-use of a society. Reconstructions and measurements of how much time people – and what kind of people – have spent and spend to get their daily requirement of water will tell much about changing material limitations for social development.

30 When the British erected the Sennar Dam on the Blue Nile in 1924, they fundamentally altered the structure of the river and the economy of this part of the Sudan; i.e., they invented structures within the structure that will influence the development of the Sudan 'for ever', but this event was not in any way embedded in the structures of the waterscape of the past.

31 See, for example, Aymard 1972: 496–7.

32 In Giddens' book *The Constitution of Society* (1984), there is no reference to or entry for nature, water, environment or the physical external world. Society is constituted totally independently of such factors. He argues that geography and sociology have come closer, because geography 'has come to contain many of the same concepts […] as sociology' (Giddens 1984: 364). Giddens' definition of contextuality is also very telling in this context. 'Contextuality means space as well as time' (Giddens 1984: 363), but both time and space are defined as social variables.

33 For a study of this, see Feldhaus 1995.

34 Typical titles include: *A River No More: The Colorado River and the West* (Fradkin 1984), *A River Lost: The Life and Death of the Columbia* (Harden 1996), *River of Sorrow. Environment and Social Control in Riparian North India 1770–1994* (Hill 1997) (20) and Richard White's *The Organic Machine: The Remaking of the Columbia River*.

35 See, for example, Williams 1980: 68. There he argues that nature is one of the most complex words in language altogether.

36 See, for example, Pratt et al. 2000; Part 1 of Jamieson 2001; Gottlieb 2003; Selin and Kalland 2003; Turner 2005; Callicott and Palmer 2005; Radkau 2008.

37 See, for example, Crosby 1972; Warren 1995; Opie 1993.

38 Glacken 1967; Worster 1985a [1977]; Pepper 1996; Coates 1998; Buttimer and Wallin 1999; Pratt and Brady 2000; Jamieson 2001; Gottlieb 2003; Selin and Kalland 2003; Turner 2005; Callicott and Palmer 2005; Radkau 2008; Allen 1963.

39 This literature is very extensive, but one book that exemplifies the point is O'Neill 1993.

40 Hughes 2006: 4. See also Simmons 1993 and Cronon 1992.
41 Catton and Dunlap 1980. This paradigm is discussed by Dickens 1992. For an early expression of this idea, see Catton 1983.
42 See Benton 2001.
43 See, for example, Barry 1999.

EPILOGUE

1   This quote is taken from Gare 1995: 58.

GLOSSARY

1   This term was first used in the TV documentary 'The Journey of Water' from 2007 and in the book from the same year, translated into English. See Tvedt 2014a.

BIBLIOGRAPHY

Aggarwal, Rimjhim M. and Tulika A. Narayan. 2004. 'Does inequality lead to greater efficiency in the use of local commons? The role of strategic investments in capacity'. *Journal of Environmental Economics and Management* 47.1: 163–82.

Aldcroft, Derek H. and Michael J. Freeman. 1983. *Transport in the Industrial Revolution*. Manchester: Manchester University Press.

Allen, Don. C. 1963. *The Legend of Noah: Renaissance, Rationalism in Art, Science, and Letters*. Urbana, IL: University of Illinois Press.

Amin, Samir. 1974. *Accumulation on a World Scale: A Critique of the Theory of Underdevelopment*. New York: Monthly Review Press.

Anthes, Rudolf. 1959. 'Egyptian theology in the Third Millennium B.C'. *Journal of Near Eastern Studies* 18.3: 169–212.

Armstrong, Adrian and Margaret Armstrong. 2006. 'A Christian perspective on water and water rights', in Terje Tvedt and Terje Oestigaard (eds), *A History of Water: The World of Water*, Vol. III, Series I (series ed. Terje Tvedt). London and New York: I.B.Tauris, 367–85.

Arnell, Nigel W. 1999. 'Climate change and global water resources'. *Global Environmental Change* 9, Suppl. 1: S31–S49.

Arnold, Philip P. and Ann Grodzins Gold. 2001. *Sacred Landscapes and Cultural Politics: Planting a Tree*. London: Ashgate.

Arsano, Yacob. 2007. 'Challenges of effective cooperation in the Nile Basin'. Paper presented at the conference *Green Wars? Environment Between Conflict and Cooperation in the Middle East and North Africa*. Organised by Heinrich Böll Foundation, Beirut 2–3 November 2007.

Aspin, Chris. 2003. *The Water-Spinners: A New Look at the Cotton Trade*. Helmshore, UK: Helmshore Local History Society.

Axelrod, Robert. 1984. *The Evolution of Cooperation*. New York: Basic Books.

Aymard, M. 1972. 'The annales and French historiography (1929–1972)'. *Journal of European Economic History* 112: 496–7.

Ayres, Robert. 2001. *Christian Baptism: A Treatise on the Mode of Administering the Ordinance by the Apostles and Their Successors in the Early Ages of the Church*. London: Charles H. Kelly.

Baartmans, Frans. 1990. *Apah, the Sacred Waters: An Analysis of a Primordial Symbol in Hindu Myths*. Delhi: R.B. Publishing.

Bagwell, Philip S. 1974. *The Transportation Revolution from 1770*. London: Batsford B.T.

Bagwell, Philip S. and Peter Lyth. 2002. *Transport in Britain, 1750–2000: From Canal Lock to Gridlock*. London: Hambledon and London.

Baines, Edward. 1835. *The History of the Cotton Manufacture in Great Britain*. London: Fisher, Fisher & Jackson.

Bairoch, Paul. 1982. 'International industrialization levels from 1750 to 1980'. *Journal of European Economic History* 11: 269–333.

Baker, Samuel. 1867. *The Albert N'yanza, Great Basin of the Nile and Explorations of the Nile Sources*, 2 vols. London: Macmillan.

———. 1884. 'Egypt's proper frontier'. *Nineteenth Century* (July): 27–46.

———. 1888. Three Articles in *The Times*, 9, 17 and 25 October.

Bakker, Karen. 2010. *Privatizing Water: Governance Failure and the World's Urban Water Crisis*. Ithaca, NY: Cornell University Press.

———. 2010. *Governance Failure and the World's Urban Water Crisis*. Ithaca, NY and London: Cornell University Press.

Bandstra, Barry L. 2009. *Reading the Old Testament: An Introduction to the Hebrew Bible*, 4th ed. Belmont, CA: Wadsworth/Cengage Learning.

Barry, John. 1999. *Rethinking Green Politics: Nature, Virtue, and Progress*. London: Sage.

Bartlett, R. V. 1990. 'Comprehensive environmental decision making, can it work?', in N. J. Vig and M. E. Kraft (eds), *Environmental Policy in the 1990s: Toward a New Agenda*. Washington, DC: CQ Press, 235–57.

Baskin, Jonathan B. and Paul J. Miranti. 1997. *A History of Corporate Finance*. Cambridge: Cambridge University Press.

Bates, Darrell. 1984. *The Fashoda Incident of 1898: Encounter on the Nile*. Oxford: Oxford University Press.

Batou, Jean. 1993. 'Nineteenth-century attempted escapes from the periphery: The cases of Egypt and Paraguay'. *Review – Fernand Braudel Center* 16.3: 279–318.

Beach, Heather L. et al. 2000. *Transboundary Freshwater Dispute Resolution: Theory, Practice, and Annotated References*. Water Resources Management and Policy Series. Tokyo and New York: United Nations University Press.

Beasley-Murray, George R. 1962. *Baptism in the New Testament*. London: Macmillan.

Beck, Ulrich. 1995. *Ecological Politics in the Age of Risk*. Cambridge: Polity Press.

———. 1999. *World Risk Society*. Cambridge: Polity Press.

Benton, T. 2001. 'Environmental sociology: Controversy and continuity', in Ann Nilsen and Terje Tvedt (eds), *Sosiologisk Tidsskrift* 1–2 (Special issue: 'Natur og amfunn. Noen Teoretiske Perspektiver'): 5–49.

Berger, Stefan, Heiko Feldner and Kevin Passmore (eds). 2003. *Writing History: Theory and Practice*. London: Hodder Education.

Bernstein, Henry T. 1960. *Steamboats on the Ganges: An Exploration in the History of India's Modernization Through Science and Technology*. Calcutta: Orient Longman.

Berkes, F. (ed.). 1989. *Common Property Resources: Ecology and Community-Based Sustainable Development*. London: Belhaven Press.

Best, Robert M. 1999. *Noah's Ark and the Ziusudra Epic*. Fort Myers, FL: Enlil Press.

Betts, Raymond F. 1972. *The Scramble for Africa; Causes and Dimensions of Empire*. Lexington, MA: Heath.

Bharati, Radhakant. 2004. *Rivers of India*. New Delhi: National Book Trust.

Bird, Elizabeth R. 1987. 'The social construction of nature: Theoretical approaches to the history of environmental problems'. *Environmental Review* 11.4: 255–64.

Blackburn, Glenn A. 1930. 'International control of the River Danube'. *Current History* 42: 1154–9.

Blair, Sheila S. and Jonathan M. Bloom. 2009. *Rivers of Paradise: Water in Islamic Art and Culture*. New Haven, CT: Yale University Press.

Blake, Nelson Manfred. 1956. *Water for the Cities: A History of the Urban Water supply in the United States*. Syracuse, NY: Syracuse University Press.

Bloch, Marc. 1935. 'Avènement et conquêtes du moulin à eau', *Annales d'histoire économique et sociale* 7: 538–63, published in English in 1967 as 'The advent and triumph of the watermill', in *Land and Work in Mediaeval Europe: Selected Papers by Marc Bloch*. Berkeley and Los Angeles: University of California Press, 136–68.

Boas, Franz. 1911. *The Mind of Primitive Man*. New York: Macmillan.

Bonell, M. 1998. 'Possible impacts of climate variability and change on tropical forest hydrology'. *Climatic Change* 39.2–3: 215–72.

Boorman, D. B. and C. E. M. Sefton. 1997. 'Recognising the uncertainty in the quantification of the effects of climate change on hydrological response'. *Climatic Change* 35.4: 415–34.

Born, S. and W. Sonzogni. 1995. 'Integrated environmental management: Strengthening the conceptualisation'. *Environmental Management* 19.2: 167–81.

Bougeant, Père. *Histoire du Traité de Westphalia ou des négociations*, Vol. 6. Paris: Didot.

Bourne, Charles B. and Patricia Wouters. 1997. *International Water Law: Selected Writings of Professor Charles B. Bourne*. London: Kluwer Law International.

Bousquet, F., M. Anderies, M. Antona, et al. 2015. 'Socio-ecological theories and empirical research: comparing social-ecological schools in action'. Research Report, Cirad-Green.

Bowden, Witt. 1925. *Industrial Society in England Towards the End of the Eighteenth Century*. New York: Macmillan.

Braudel, Fernand. 1975. *The Mediterranean and the Mediterranean World in the Age of Philip II*. Trans. Siân Reynolds. New York: Harper Torchbooks.

———. 1979. *The Structures of Everyday Life, Civilization and Capitalism, 15th–18th Century*. New York: Harper Row.

———. 1980. *On History*. Chicago: University of Chicago Press.

———. 1990. *The Identity of France. Vol. II: People and Production*. London: Fontana Press.

Bridgeman, Howard A. and John E. Oliver. 2006. *The Global Climatic System: Patterns, Processes and Teleconnections*. Cambridge: Cambridge University Press.

Broadberry, Steven. 2007. *Recent Developments in the History and Theory of Very Long Run Growth*. The Warwick Economics Research Paper Series (TWERPS) 818, Department of Economics, University of Warwick.

Brown, Michael E. 1998. *Theories of War and Peace: An International Security Reader*. Cambridge, MA: MIT Press.

Brown, Robert Glenn. 1970. *Fashoda Reconsidered: The Impact of Domestic Politics on French Policy in Africa 1893–1898*. Baltimore, MD: Johns Hopkins Press.

Bryer, R. A. 1999. 'Marx and accounting'. *Critical Perspectives on Accounting* 10.5: 683–709.

———. 2000. 'The history of accounting and the transition to capitalism in England'. Part 1: 'Theory'. *Accounting, Organizations and Society* 25.2: 131–62.

Budge, E. A. Wallis. 1989 [1923]. *The Book of the Dead*. London: Arkana.

Budyko, M. I. 1982. *The Earth's Climate: Past and Future*. International Geophysics Series, 29. New York: Academic Press.

Burroughs, William James. 2007. *Climate Change: A Multidisciplinary Approach*. Cambridge: Cambridge University Press.

Buttimer, A. and L. Wallin. 1999. *Nature and Identity in Cross-Cultural Perspective*. Dordrecht, Netherlands: Kluwer Academic Press.

Butzer, Karl. 1976. *Early Hydraulic Civilizations in Egypt: A Study in Cultural Ecology*. Chicago and London: University of Chicago Press.

Buzan, Barry, Charles A. Jones and Richard Little. 1993. *The Logic of Anarchy: Neorealism to Structural Realism*. New York: Columbia University Press.

Cain, P. J. and A. G. Hopkins. 2002. *British Imperialism, 1688–2000*. London: Longman.

Callicott, J. Baird and Clare Palmer. 2005. *Environmental Philosophy: Critical Concepts in the Environment, Society and Politics*, Vol. 2. London: Routledge.

Callicott, J. Baird and Roger T. Ames (eds). 1989. *Nature in Asian Traditions of Thought: Essays in Environmental Philosophy*. Albany, NY: State University of New York Press.

Capon, S. J. et al. 2013. 'Riparian zones in the 21st century: hotspots for climate change adaptation?'. *Ecosystems* 16: 359–81.

Carter, Harold. 1981. *The Study of Urban Geography*. London: Edward Arnold.

———. 1995. *The Study of Urban Geography*. London: Edward Arnold.

Castleden, Rodney. 1998. *Atlantis Destroyed*. London: Routledge.

Catton, William R. Jr. 1983. 'Need for a new paradigm'. *Sociological Perspectives* 26.1: 3–15.

Catton, William R. Jr. and Riley E. Dunlap. 1980. 'A new ecological paradigm for post-exuberant sociology'. *American Behavioral Scientist* 24.1: 15–47.

Chamberlain, Joseph P. 1923. *The Regime of the International Rivers: Danube and Rhine.* Studies in History, Economics, and Public Law, V 105, No 1. New York: AMS Press.

Chao, Kang. 1977. *The Development of Cotton Textile Production in China.* London: Harvard University Press.

Châtel, Francesca de. 2006. 'Bathing in divine waters: Water and purity in Judaism and Islam', in Terje Tvedt and Terje Oestigaard (eds), *A History of Water: Ideas of Water from Ancient Societies to the Modern World*, Vol. I, Series II (series ed. Terje Tvedt). London and New York: I.B.Tauris, 273–98.

Chélu, A. 1891. *De l'équateur à la Méditerranée: Le Nil, le Soudan, l'Égypt.* Paris: Chaix.

Chitwood, A. 2004. *Death by Philosophy: The Biographical Tradition in the Life and Death of the Archaic Philosophers Empedocles, Heraclitus, and Democritus.* Ann Arbor, MI: University of Michigan Press, 273–98.

Cioc, Mark. 2002. *The Rhine: An Eco-Biography, 1815–2000.* Seattle: University of Washington Press.

Ciriacono, Salvatore. 2006. *Building on Water: Venice, Holland and the Construction of the European Landscape in the Early Modern Times.* Oxford and New York: Berghahn.

Clough, Shepard B. and Charles W. Cole. 1946. *Economic History of Europe.* Boston, MA: Heath.

Coates, P. 1998. *Nature: Western Attitudes Since Ancient Times.* Cambridge: Polity Press.

Cocheris, Jules. 1903. *Situation internationale de l'Egypte et du Soudan.* Paris: Plon-Nourrit et Cie.

Cochrane, Allan. 2005. 'Cities', in P. Daniels, M. Bradshaw, et al. (eds), *An Introduction to Human Geography.* Edinburgh: Pearson, 213–29.

Cohn, Norman. 1996. *Noah's Flood: The Genesis Story in Western Thought.* New Haven, CT and London: Yale University Press.

Collingwood, R. G. and T. M. Knox. 1946. *The Idea of History.* Oxford: Oxford University Press.

———. 1999. *The Principles of History: And Other Writings in Philosophy of History*, ed. W. H. Dray and Jan Van der Dussen. New York: Clarendon Press.

———. 2014 [1945]. *The Idea of Nature.* Eastford, CT: Martino Fine Books.

Collins, Robert O. (ed.). 1968a. *Problems in African History.* Englewood Cliffs, NJ: Prentice Hall.

———. 1968b. *King Leopold, England and the Upper Nile 1899–1909.* New Haven, CT: Yale University Press.

———. (ed.). 1969. *The Partition of Africa: Illusion or Necessity.* New York: John Wiley.

———. 1971. *Land Beyond the Rivers. The Southern Sudan, 1898–1918.* New Haven, CT: Yale University Press.

———. 1983. *Shadows in the Grass, Britain in the Southern Sudan, 1918–1956*, New Haven, CT: Yale University Press.

———. 1996. *The Waters of the Nile. Hydropolitics and the Jonglei Canal 1900–1988.* Princeton, NJ: Markus Weiner.

Crafts, N. F. R. 1977. 'Industrial Revolution in England and France: Some thoughts on the question: "Why was England first?"'. *Economic History Review* 30: 429–41.

———. 1985. *British Economic Growth During the Industrial Revolution.* New York: Oxford University Press.

Cromer, Earl of (Evelyn Baring). 1885–1907. *Reports by His Majesty's Agent and Consul-General on the Finances, Administration and Conditions of Egypt and the Sudan* (annual reports). London: MSO.

———. 1908. *Modern Egypt*, 2 vols. London: Macmillan.

Cronon, William. 1992. 'A place for stories: Nature, history, and narrative'. *The Journal of American History* 78.4: 1347–76.

Crosby, Alfred W. Jr. 1972. *The Columbian Exchange: Biological and Cultural Consequences of 1492*. Westport, CT: Greenwood Press.

Crouchley, A. E. 1938. *The Economic Development of Modern Egypt*. London: Longmans Green.

Crowfort, G. M. 1924. 'The handspinning of cotton'. *Sudan Notes & Records* 7.2: 83–9.

Croxton, Derek and Anuschka Tischer. 2002. *The Peace of Westphalia: A Historical Dictionary*. Westport, CT: Greenwood.

Cullet, Philippe et al. 2010. *Water Law for the Twenty-First Century*. London: Routledge.

Cuvier, Georges. 1864. *Discours sur les révolutions du globe*, ed. Jean C. F. Hoefer. Paris: Firmin Didot.

Dai, Aiguo, K. Trenberth and T. Qian. 2004. 'A global dataset of Palmer Drought Severity Index for 1870–2002: Relationship with soil moisture and effects of surface warming'. *Journal of Hydrometeorology* 5.6: 1117–30.

Dalby, Simon. 1998. 'Ecological metaphors of security: World politics in the biosphere'. *Alternatives: Global, Local, Political* 23.3: 291–319.

Daniell, William and Richard Ayton. 1814. *A Voyage Round Great Britain*. London: Longman.

Darian, Steven G. 1978. *The Ganges in Myth and History*. Honolulu: University Press of Hawaii.

Davies, Jonathan and David L. Imbroscio (eds). 2009. *Theories of Urban Politics*. London: Sage.

Davis, Mike. 2006. *City of Quartz: Excavating the Future in Los Angeles*. New York: Verso.

Deane, Phyllis. 1979. *The First Industrial Revolution*, 2nd ed. Cambridge: Cambridge University Press.

Dellapenna, J., and Joyeeta Gupta. 2014. '2014', in Terje Tvedt, Owen Roberts and Tadesse Kassa (eds), *A History of Water: Sovereignty and International Water Law*, Vol. II, Series III (series ed. Terje Tvedt). London and New York: I.B.Tauris, 27–47.

Delli Priscoli, Jerome. 1990. *Public Involvement; Conflict Management; and Dispute Resolution in Water Resources and Environmental Decision Making*. Ft. Belvoir, VA: U.S. Army Corps of Engineers.

Dharampal, Shri (ed.). 1971. *Indian Science and Technology in the Eighteenth Century*. Goa: Other India Press.

Diamond, Jared. 1997. *Guns, Germs, and Steel: The Fates of Human Societies*. New York, NY: W. W. Norton.

———. 2005. *Collapse: How Societies Choose to Fail or Succeed*. New York: Viking Books.

Dickens, Peter. 1992. *Society and Nature: Towards a Green Social Theory*. Philadelphia: Temple University Press.

Dickinson, Robert E. 2003. 'Overview: The climate system', in Thomas D. Potter and Bradley Colman (eds), *Handbook of Weather, Climate, and Water: Atmospheric Chemistry, Hydrology, and Societal Impacts*. London: Wiley-Interscience, 119–29.

Diefendorf, J. M., and K. Dorsey (eds). 2005. *City, Country, Empire: Landscapes in Environmental History*. Pittsburgh, PA: University of Pittsburgh Press.

Dinar, Shlomi, 2008. *International Water Treaties: Negotiation and Cooperation along Transboundary Rivers*. London: Routledge.

Dodgen, Randall A. 2001. *Controlling the Dragon, Confucian Engineers and the Yellow River in Late Imperial China*. Honolulu: University of Hawaii Press.

Doniger, Wendy. 1998. *The Implied Spider: Politics and Theology in Myth*. New York: Columbia University Press.

———. 2010. 'Flood myths', in Terje Tvedt and Terje Oestigaard (eds), *A History of Water: Ideas of Water: From Ancient Societies to the Modern World*, Vol. II, Series II (series ed. Terje Tvedt). London and New York: I.B.Tauris, 424–40.

Douglas, Mary. 1994. *Purity and Danger: An Analysis of Concept of Pollution and Taboo*. London: Routledge.

Doyle, Barry M. 2009. 'A decade of urban history: Ashgate's Historical Urban Series'. *Urban History* 36.3: 498–512.

Dufournaud, C. and J. Harrington. 1990. 'Temporal and spatial distribution of benefit and costs in river-basin schemes: A cooperative game approach'. *Environmental Planning A* 22.5: 615–28.

Dundes, Allan (ed.). 1988. *The Flood Myth*. Berkeley, CA: University of California Press

Dupuis, C. E., 1904. 'Report upon Lake Tana and the rivers of the Eastern Soudan', attached to William E. Garstin. *Report upon the Basin of the Upper Nile with Proposals for the Improvement of that River*. Cairo: Ministry of Public Works.

Durkheim, Émile. 1925. *The Elementary Forms of the Religious Life: A Study in Religious Sociology*, trans. Joseph W. Swain. London/New York: George Allen & Unwin/Macmillan.

———. 1960a. 'Sociology and its scientific field', in Kurt H. Wolff (ed.), *Émile Durkheim, 1858–1917: A Collection of Essays with Translations and Biography*. Columbus, OH: Ohio State University Press, 354–63.

———. 1960b. 'Prefaces to L'Année Sociologique, 1960', in Kurt H. Wolff (ed.), *Émile Durkheim, 1858–1917: A Collection of Essays with Translations and Biography*. Columbus, OH: Ohio State University Press, 341–53.

———. 1960c. 'Sociology', in Kurt H. Wolff, (ed.), *Emile Durkheim, 1858–1917: A Collection of Essays with Translations and Biography*. Columbus: Ohio State University Press, 376–85.

———. 1966 [1904]. *The Rules of Sociological Method*. New York: Free Press.

———. 1984 [1893]. *The Division of Labor in Society*. New York: Free Press.

Dyrvik, Ståle, Anders Bjarne Fossen, Tore Grønlie, et al. 1979. *Norsk Økonomisk Historie 1500–1970*, vol. 1, *1500–1850*. Oslo: Universitetsforlaget.

Eade, John and Christopher Mele (eds). 2002. *Understanding the City: Contemporary and Future Perspectives*. Oxford: Blackwell.

Eagleson, Peter. S. 1978. 'Climate, soil, and vegetation 6. Dynamics of the annual water balance'. *Water Resources Research* 14.5: 449–76.

Eastman, Lloyd E. 1988. *Family, Fields and Ancestors: Constancy and Change in China's Social and Economic History, 1550–1849*. Oxford: Oxford University Press.

Eck, Diana. L. 1983. *Banaras, City of Light*. New Delhi: Penguin Books.

Eddy, Frank W. 1991. *Archaeology: A Cultural-Evolutionary Approach*, 2nd ed. Englewood Cliffs, NJ: Prentice Hall.

Eliade, Mircea. 1979. *The Sacred and the Profane*. New York: Harcourt Brace Jovanovich.

Ellison, Thomas. 1968. *The Cotton Trade of Great Britain*. New York: A. M. Kelley.

Elvin, Mark. 1972. 'The high-level equilibrium trap: The causes of the decline of invention in the traditional Chinese textile industries', in W. E. Willmott (ed.), *Economic Organization in Chinese Society*. Stanford, CA: Stanford University Press, 137–72.

———. 1973. *The Pattern of the Chinese Past*. London: Eyre Methuen.

———. 1977. 'Market towns and waterways: The County of Shang-hai from 1480 to 1910', in G. William Skinner (ed.), *The City in Late Imperial China*. Stanford, CA: Stanford University Press, 441–75.

———. 1998. 'Unseen lives: The emotions of everyday existence mirrored in Chinese popular poetry of the mid-seventeenth to the mid-nineteenth century', in Roger T. Ames et al. (eds), *Self as Image in Asian Theory and Practice*. Albany, NY: State University of New York Press, 113–200.

Emanuel, William R., H. H. Shugart and M. P. Stevenson. 1985. 'Climatic change and the broad-scale distribution of terrestrial ecosystem complexes'. *Climate Change* 7.1: 29–43.

Emin Pasha. 1879. 'Strombarren des Bahr el-Gebel'. *Petermanns Mitteilungen* 879.

Enfield, David. B., 1989. 'El Niño, past and present'. *Reviews of Geophysics* 27.1: 159–87.

Engel, J. Ronald and Joan G. Engel (eds). 1990. *Ethics of Environment and Development: Global Challenge and the International Response*. London: Belhaven.

Ewing, Bradley T., Jerry S. Rawls and Jamie B. Kruse (eds). 2005. *Economics and the Wind*. New York: Nova Science.

Exline, Christopher H., Gary L. Peters and Robert P. Larkin. 1982. *The City: Patterns and Processes in the Urban Ecosystem*. Boulder, CO: Westview Press.

Fagan, Brian, 2011. *Elixir: A History of Water and Humankind*. New York: Bloomsbury.

Falk, Richard. 1969. 'The interplay of Westphalia and charter conceptions of the international legal order', in R. Falk and C. Black (eds), *The Future of the International Legal Order, Vol. 1, Trends and Patterns*. Princeton, NJ: Princeton University Press, 32–70.

Falkenmark, Malin and Jan Lundqvist. 1999. 'Focusing on the upstream/downstream interdependencies and conflicts of interests – steps and procedures towards coping with management challenges'. *Analytical Summary from the Proceedings of the SIWI/ IWRA Seminar*, Stockholm, Sweden.

Faruqui, Naser, Asit K. Biswas and Murad J. Bino. 2001. *Water Management in Islam*, Water Resources Management and Policy series. Tokyo and New York: United Nations University Press.

Feldhaus, Anne. 1995. *Water and Womanhood: Religious Meanings of Rivers in Maharashtra*. Oxford: Oxford University Press.

Ferguson, Niall. 2011. *Civilization: The West and the Rest*. New York: Penguin Press.

Fitzmaurice, M., Olufemi Elias and A. V. Lowe. 2004. *Watercourse Co-Operation in Northern Europe: A Model for the Future*. The Hague: TMC Asser Press.

Florescano, Enrique. 2002. *The Myth of Quetzalcoatl*, trans. Lysa Hochroth. New York: Johns Hopkins University Press.

Florescano, Enrique and Raul Velazquez. 2002. *National Narratives in Mexico: A History. Norman*. OK: University of Oklahoma Press.

Flynn, Dennis and Arturo Giráldez. 1995. 'Born with a "silver spoon": The origin of world trade in 1571'. *Journal of World History* 6.2: 201–22.

Fogel, Joshua A. 1988. 'The debates over the Asiatic mode of production in Soviet Russia, China and Japan'. *American Historical Review* 93.1: 56–79.

Fogel, Robert W. 1983. 'Scientific history and traditional history', in R. W. Fogel and G. R. Elton (eds), *Which Road to the Past?: Two Views of History*. New Haven, CT: Yale University Press, 23–34.

Forbes, Robert J. 1956. 'Power', in Charles Singer, Eric John Holmyard, Alfred Rupert Hall et al., *A History of Technology, II*. London: Oxford University Press, 601–6.

Forrest, Charles Rasmus. 1824. *A Picturesque Tour Along the Ganges and Jumna in India Consisting of Twenty-Four Highly Finished and Coloured Views, A Map and Vignettes from Original Drawings Made on the Spot*. London: R. Ackermann.

Fradkin, Philip. 1996. *A River No More: The Colorado River and the West*. Berkeley, CA: University of California Press.

Frank, André Gunder. 1969. *Capitalism and Underdevelopment in Latin America: Historical Studies of Chile and Brazil*. New York: Monthly Review Press.

Frankfort, Henri. 1948. *Kingship and the Gods. A Study of Ancient Near Eastern Religion as the Integration of Society and Nature*. Chicago: University of Chicago Press.

Frazer, James George. 1922. *The Golden Bough: A Study in Magic and Religion*. New York: Macmillan.

Frederick, Kenneth D. and David C. Major. 1997. 'Climate change and water resources'. *Climatic Change* 37.1: 7–23.

Frey, Frederick W. 1993. 'The political context of conflict and cooperation over international river basins'. *Water International* 18.1: 54–68.

Frye, Northrop and Jay Macpherson. 2004. *Biblical and Classical Myths: The Mythological Framework of Western Culture*. Toronto: University of Toronto Press.

Fu Zehong. 1995 [1725]. *Xing Shui Jin Jian (Golden Mirror of the Flowing Waters)*, in Colin A. Ronan, *The Shorter Science and Civilization in China*, Vol. 5. Cambridge: Cambridge University Press.

Gallagher, John and Ronald Robinson. 1953. 'The imperialism of free trade'. *The Economic History Review*, Second Series VI(I): 1–15.

Gandy, Matthew. 2003. *Concrete and Clay: Reworking Nature in New York City*. Boston, MA: MIT Press.

Garbarino, Merwyn S. 1977. *Sociocultural Theory in Anthropology: A Short History*. Illinois: Waveland Press.

Gare, A. 1995. *Postmodernism and the Environmental Crisis*. New York: Routledge.

Garstin, William E. 1899. *Note on the Soudan*. Cairo: Ministry of Public Works.

———. 1900–6. 'Report upon the Administration of the Public Works Department for 1899–1905. With reports by the officers in charge of the several branches of the Administration', 7 vols, Cairo.

———. 1901. *Report as to Irrigation Projects on the Upper Nile*, in Foreign Office, Blue Book No. 2, 1901 in Despatch from His Majesty's Agent and Consul-General, Cairo.

———. 1904. *Report upon the Basin of the Upper Nile with Proposals for the Improvement of that River*. Cairo: Ministry of Public Works.

———. 1905. 'Some problems of the Upper Nile. *The Nineteenth Century and After* VLII.343: 345–66.

———. 1909. 'Fifty years of Nile exploration and some of its results'. *The Geographical Journal* 33.2: 117–52.

Geiger, Rudolph. 1950. *The Climate Near the Ground*, 2nd ed. Cambridge, MA: Harvard University Press.

Gerschenkron, Alexander. 1962. *Economic Backwardness in Historical Perspective: A Book of Essays*. Cambridge, MA: Harvard University Press.

Gerten, Dieter. 2006. 'How water transcends religions and epochs: Hydrolatry in early European religions and Christian syncretism', in Terje Tvedt and Terje Oestigaard (eds), *A History of Water: Ideas of Water from Ancient Societies to the Modern World*, Vol. I, Series II (series ed. Terje Tvedt). London and New York: I.B.Tauris, 323–43.

Giddens, Anthony. 1971. *Capitalism and Modern Social Theory: An Analysis of the Writings of Marx, Durkheim and Max Weber*. Cambridge: Cambridge University Press.

———. 1984. *The Constitution of Society: Outline of the Theory of Structuration*. Berkeley, CA: University of California Press.

———. 1986. *Durkheim*. London: HarperCollins.

———. 1989. *Sociology*. Cambridge: Polity Press.

———. 1992. *Human Societies: An Introductory Reader in Sociology*. Cambridge: Polity Press.

Giddens, Anthony and Christopher Pierson. 1998. *Conversations with Anthony Giddens: Making Sense of Modernity*. Stanford, CA: Stanford University Press.

Gimpel, Jean. 1988. *The Medieval Machine: The Industrial Revolution of the Middle Ages*. Aldershot, UK: Wildwood House.

Glacken, Clarence J. 1967. *Traces on the Rhodian Shore: Nature and Culture in Western Thought from Ancient Times to the End of the Eighteenth Century*. Berkeley, CA: University of California Press.

Gleichen, A. E. W. (ed.). 1905. *The Anglo-Egyptian Sudan*, 2 vols. London.

Goody, R. M., and Y. L. Yung. 1989. *Atmospheric Radiation: Theoretical Basis*. Oxford: Oxford University Press.

Gordon, Line J., Will Steffen, Bror F. Jönsson et al. 2005. 'Human modification of global water vapor flows from the land surface'. *Proceedings of the National Academy of Sciences of the United States of America* 102.21: 7612–17.

Gordon, Stuart. 1996. *The Book of Miracles: From Lazarus to Lourdes*. London: Headline Book Publishing.

Gore, Al. 2007. Nobel Lecture, Oslo City Hall, 10 December 2007. Available at: http://www.nobelprize.org/nobel_prizes/peace/laureates/2007/gore-lecture_en.html. (Accessed 13 June 2014.)

Gottdiener, Mark. 1994. *The New Urban Sociology*. London: McGraw-Hill.

Gottdiener, Mark and Leslie Budd. 2005. *Key Concepts in Urban Studies*. London: Sage.

Gottlieb, Roger S. (ed.). 2003. *This Sacred Earth: Religion, Nature, Environment*. London: Routledge.

Green, Alberto R. W. 2003. *The Storm-God in the Ancient Near East*. Biblical and Judaic Studies, Vol. 8. Winona Lake, IN: Eisenbrauns.

Guillerme, André E. 1988 [1983]. *The Age of Water: The Urban Environment in the North of France, A.D. 300–1800*. College Station, TX: Texas A&M University Press.

Gunderson, L. H., and C. S. Holling (eds). 2002. *Panarchy: Understanding Transformations in Human and Natural Systems*. Washington, DC: Island Press.

Habermas, Jürgen. 1979. *Communication and the Evolution of Society*. London: Heinemann.

Hall, Peter Geoffrey. 1998. *Cities in Civilization: Culture, Innovation and Urban Order*. London: Weidenfeld & Nicolson.

———. 2002. *Cities of Tomorrow: An Intellectual History of Urban Planning and Design in The Twentieth Century*. Oxford: Basil Blackwell.

Hall, Tim. 1998. *Urban Geography*. London: Routledge.

Hall, Tim and Heather Barrett. 2012. *Urban Geography*. Routledge Contemporary Human Geography Series. London: Routledge.

Haraway, Donna. 1991. *Simians, Cyborgs and Women: The Reinvention of Nature*. London: Free Association Books.

Harden, Blaine. 1996. *A River Lost: The Life and Death of the Columbia*. New York: W. W: Norton.

Hardin, Garret. 1968. 'The tragedy of the commons'. *Science* 162: 1243–8.

Harding, Christopher and C. L. Lim (eds). 1999. *Renegotiating Westphalia: Essays and Commentary on the European and Conceptual Foundations of Modern International Law*. The Hague: Martinus Nijhoff Publishers.

Hargrove, Eugene C. 1986. *Religion and Environmental Crisis*. Athens, GA: University of Georgia Press.

Harley, Knick C. 1982. 'British industrialization before 1841: Evidence of slower growth during the Industrial Revolution'. *Journal of Economic History* 42.2: 267–89.

Harper, Michael. 1970. *The Baptism of Fire*. London: Fountain Trust.

Harris, Ruth. 1999. *Lourdes: Body and Spirit in the Secular Age*. New York: Viking.

Hart, Henry C. 1956. *New India's Rivers*. Bombay: Orient Longmans.

Hartmann, Dennis L. 1994. *Global Physical Climatology*. Salt Lake City, UT: Academic Press.

Hartshorn, Truman Asa. 1980. *Interpreting the City: An Urban Geography*. New York and Chichester: Wiley.

Hartwell, R. M. 1967. *The Causes of the Industrial Revolution in England*. London: Methuen.

Harvey, David. 1996a. *Justice, Nature and the Geography of Difference*. Cambridge: Blackwell.

———. 1996b. 'Cities or urbanization'. *City* 1(1–2): 38–61.

Hassan, Fekri. 1997. 'The dynamics of a riverine civilization: a geoarchaeological perspective on the Nile Valley, Egypt'. *World Archeology* 29.1 (Riverine Archaeology): 51–74.

Held, David. 1995. *Democracy and the Global Order: From the Modern State to Cosmopolitan Governance*. Cambridge: Polity Press.

Held, David. 2002. 'Law of States, Law of Peoples: three models of sovereignty'. *Legal Theory* 8.2: 1–44.

Held, Isaac M. and Brian J. Soden. 2000. 'Water vapor feedback and global warming'. *Annual Review Energy Environment* 25: 441–75.

Herbert, David T. 1968. *Urban Geography: A Social Perspective*. Newton Abbot: David and Charles.

Herbert, David T. and Colin J. Thomas. 1982. *Urban Geography: A First Approach*. New York: John Wiley.

271

Herodotus. 1960. *The Histories*. London: Penguin Books.

Hertz, Robert. 1996. *Sin and Expiation in Primitive Societies*. Occasional Papers No. 2. British Centre for Durkheimian Studies, Oxford.

Hill, Christopher V. 1997. *River of Sorrow: Environment and Social Control in Riparian North India 1770–1994*. Monograph and Occasional Paper Series, No. 55. Association for Asian Studies.

Hills, Richard. L. 1970. *Power in the Industrial Revolution*. Manchester: Manchester University Press.

Hippocrates. 1923. 'Airs water places', in *Hippocrates*. The Loeb Classical Library. London and Cambridge, MA: Heinemann and Harvard University Press, 65–138.

Hobsbawm, Eric. 1994. *The Age of Extremes: The Short Twentieth Century, 1914–1991*. London: Michael Joseph.

Holm, Jean and John Bowker (eds). 1994. *Attitudes to Nature*. Themes in Religious Studies Series. London: Francis Pinter.

Holmsen, Andreas. 1961. *Norges historie fra de eldste tider til 1660*. Oslo: Universitetsforlaget.

Holt, P. M. 1967. *A Modern History of the Sudan: From the Funj Sultanate to the Present Day*, 3rd ed. London: Weidenfeld and Nicolson.

Homer-Dixon, Thomas F. 1999. *Environment, Scarcity and Violence*. Princeton, NJ: Princeton University Press.

Hosni, Sayed Muhamed. 1975. 'Legal problems of the development of the River Nile. Ph.D. Thesis. Ann Arbor, MI: University Microfilms.

Howarth, William. 2014. 'The history of water law in the common law tradition', in Terje Tvedt, Owen Roberts and Tadesse Kassa (eds), *A History of Water: Sovereignty and International Water Law*, Vol II, Series III (series ed. Terje Tvedt). London and New York: I.B.Tauris, 66–105.

Hughes, J. Donald. 2006. *What Is Environmental History?* Cambridge: Polity Press.

Huntington, Ellsworth. 1945. *Mainsprings of Civilization*. New York: Wiley.

Huntington, Ellsworth and Sumner W. Cushing. 1922. *Principles of Human Geography*. New York: John Wiley.

Huntington, Ellsworth and Stephen S. Visher. 1922: *Climatic Changes, Their Nature and Causes*. New Haven, CT: Yale University Press.

Huntington, Thomas G. 2005. 'Evidence for intensification of the global water cycle: Review and synthesis'. *Journal of Hydrology* 319.1–4: 83–95.

Hurst, H. E. 1927. 'Progress in the study of the hydrology of the Nile in the last twenty years'. *The Geographical Journal* 70.5: 440–63.

Hurst, H. E. and D. A. F. Watt. 1928. 'The measurement of the discharge of the Nile through the sluices of the Aswan Dam. Final Conclusions and Tables of Results', Ministry of Public Works, Egypt, Physical Department. *Physical Department Paper*, no. 24. Cairo: Government Press.

Iggers, Georg G. and Q. Edward Wang. 2008. *A Global History of Modern Historiography*. Harlow, UK: Pearson Education.

Imsen, Steinar and Harald Winge. 1999. *Norsk Historisk Leksikon*. Oslo: Cappelen Akademisk.

Intergovernmental Panel on Climate Change. 1998. *The Regional Impacts of Climate Change: An Assessment of Vulnerability*. Cambridge: Cambridge University Press.

Issawi, Charles Philip. 1982. *An Economic History of the Middle East and North Africa*. London: Methuen.

Jackman, William T. 1916. *The Development of Transportation in Modern England*. Cambridge: Cambridge University Press.

Jackson, Robert. 2007. *Sovereignty: The Evolution of an Idea*. Cambridge: Polity Press.

Jacobsen, Trudy, C. J. G. Sampford and Ramesh Chandra Thakur. 2008. *Re-Envisioning Sovereignty: The End of Westphalia?* Law, Ethics and Governance Series. Aldershot, UK: Ashgate.

Jain, Sharad K, Pushpendra K. Agarwal and Vijay P. Singh. 2007. *Hydrology and Water Resources of India*. Dordrecht, Netherlands: Springer Verlag.

Jamieson, Dale. 2001. *A Companion to Environmental Philosophy*. Blackwell Companions to Philosophy. Malden, MA: Blackwell.

Jin Fu. 1995 [1689, 1767]. *Zhi He Fang Lue (Methods of River Control)*, in Colin A. Ronan, *The Shorter Science and Civilization in China*, vol. 5. Cambridge: Cambridge University Press.

Johnson, Robert. 2003. *British Imperialism: Histories and Controversies*. New York: Palgrave Macmillan.

Johnson, R. G. 1997. 'Climate control requires a dam at the Strait of Gibraltar'. *Eos Transactions, American Geophysical Union* 78.27: 277, 280–1.

Johnston, R. J. 1980. *City and Society: An Outline for an Urban Geography*. London: Hutchinson.

———. 1992. 'Laws, states, and super-states: International law and the environment'. *Applied Geography* 12.3: 211–28.

Johnston, R. J., Derek Gregory, Geraldine Pratt, et al. (eds). 2008. *Dictionary of Human Geography*. Oxford: Blackwell.

Jones, Evan T. 2000. 'River navigation in medieval England'. *Journal of Historical Geography* 26.1: 60–82.

Kabat, Pavel, Henk van Schaik, et al. 2003. 'Climate changes the water rules: How water managers can cope with today's climate variability and tomorrow's climate change'. Dialogue on Water and Climate: The Netherlands.

Kabat, Pavel, Martin Claussen, Stacey Whitlock et al. 2013. *Vegetation, Water, Humans and the Climate: A New Perspective on an Interactive System*. Berlin: Springer Berlin.

Kaeckenbeeck, Georges. 1920. *International Rivers*. London: HMSO.

Kaika, Maria. 2003. 'Constructing scarcity and sensationalizing water politics'. *Antipode* 35: 919–54.

———. 2005. *City of Flows: Modernity, Nature, and the City*. New York: Routledge.

Kalma, J. D. and M. Sivapalan (eds). 1995. *Scale Issues in Hydrological Modelling*. Chichester, UK: Wiley.

Kampmann, Christoph. 2010. 'Peace impossible? The Holy Roman Empire and the European state system in the seventeenth century', in Olaf Asbach and Peter Schröder (eds), *War, the State and International Law in Seventeenth-Century Europe*. London: Ashgate, 197–211.

Kanefsky, John William. 1979. 'The diffusion of power technology in British industry, 1760–1870. Ph.D. thesis, University of Exeter.

Kang Jitian. 1995 [1804]. *He Qu Ji Wen (Notes on Rivers and Canals)*, in Colin A. Ronan, *The Shorter Science and Civilization in China*, Vol. 5. Cambridge: Cambridge University Press.

Kaplan, David, James O. Wheeler and Steven Holloway. 2008. *Urban Geography*. Hoboken, NJ: Wiley.

Keiter, Robert B. 1994. 'Beyond the boundary line: Constructing a law of ecosystem management'. *University of Colorado Law Review* 65.2: 293–333.

Kelman, Ari. 2003. *A River and Its City: The Nature of Landscapes in New Orleans*. Berkeley, CA: University of California Press.

Keohane, Robert and Joseph S. Nye. 1972. *Transnational Relations and World Politics*. Cambridge, MA: Harvard University Press.

Keohane, Robert and Joseph S. Nye. 1977. *Power and Interdependence: World Politics in Transition*. Boston, MA: Little, Brown.

Khan, M. Abbas. 2005. *Encyclopedia of Indian Geography*, Vol. 2. New Delhi: Anmol Publications.

Khasandi-Telewa, Vicky I. 2006. 'Of frogs' eyes and cows' drinking water: water and folklore in western Kenya', in Terje Tvedt and Terje Oestigaard (eds). *A History of*

Water: The World of Water, Vol. III, Series I, (series ed. Terje Tvedt). London and New York: I.B.Tauris, 289–310.

Kiehl, J. T. and Kevin E. Trenberth. 1997. 'Earth's annual global mean energy budget'. *Bulletin of the American Meteorological Society* 78.2: 197–208.

Knox, Paul and Linda M. McCarthy. 2012. *Urbanization: An Introduction to Urban Geography*, 3rd ed. Chicago: Chicago University press.

Knox, Paul and Steven Pinch. 2009. *Urban Social Geography: An Introduction*. New York: Pearson Prentice Hall.

Kodiyanplakkal, Joisea Joseph. 2006. 'River cult and water management practices in ancient India', in Terje Tvedt and Terje Oestigaard (eds), *A History of Water: The World of Water*, Vol. III Series I (series ed. Terje Tvedt). London and New York: I.B.Tauris, 385–405.

Koutsoyiannis, Demetris and Anna Patrikiou. 2015. 'Water control in ancient Greek cities', in Terje Tvedt and Terje Oestigaard, *A History of Water: Water and Urbanization*, Vol. I, Series III (series ed. Terje Tvedt). London and New York: I.B.Tauris, 130–49.

Krader, Lawrence and M. M. Kovalevskii. 1975. *The Asiatic Mode of Production: Sources, Development and Critique in the Writings of Karl Marx*. Assen, Netherlands: Van Gorcum.

Kramer, Samuel N. and John Maier. 1989. *Myths of Enki, the Crafty God*. Oxford: Oxford University Press.

Kurita, Y. 1989. 'The concept of nationalism in the White Flag League Movement', in Mahasin Abdelgadir Hag al Safi, *The Nationalist Movement in the Sudan*. Sudan Library Series 15. Khartoum: Institute of African and Asian Studies, University of Khartoum, 14–62.

Lambert, John Chisholm. 1903. *The Sacraments in the New Testament: Being the Kerr Lectures for 1903*. Edinburgh: T&T Clark.

Lambert, Peter and Phillipp R. Schofield. 2004. *Making History: An Introduction to the History and Practices of a Discipline: An Introduction to the Practices of History*. London: Routledge.

Landes, David, 1991. 'Does it pay to be late?', in Jean Batou, *Entre développement et sous-développement: les tentatives précoces d'industrialisation de la périphérie, 1800–1870 (Between Development and Underdevelopment: The Precocious Attempts at Industrialization of the Periphery, 1800–1870)*. Geneva: Libr. Droz, 43–67.

———. 1998. *The Wealth and Poverty of Nations*. London: Abacus.

Langer, William J. 1936. *The Diplomacy of Imperialism, 1890–1902*, 2 vols. New York: Knop.

Latour, Bruno. 1993. *We Have Never Been Modern*. Cambridge, MA: Harvard University Press.

———. 1999. *Pandora's Hope: Essays on the Reality of Science Studies*. Cambridge, MA: Harvard University Press.

Lawson, F. H. and Bernard Rudden. 2002. *The Law of Property*. 3rd ed. Oxford: Oxford University Press.

Leach, Edmund. 1969. *Genesis as Myth and Other Essays*. London: Jonathan Cape.

Lees, Andrew and Lynn Hollen Lees. 2007. *Cities and the Making of Europe, 1750–1914*. Cambridge: Cambridge University Press.

Lefebvre, Henri. 1991 [1974]. *The Production of Space*. Cambridge: Blackwell.

Lehtonen, Tommi. 1999. *Punishment, Atonement and Merit in Modern Philosophy of Religion*. Helsinki: Luther-Agricola Society.

Leimu, Pekka. 2002. 'The Finnish sauna and its Finnishness', in Susan C. Anderson and Bruce H. Tabb, *Water, Leisure and Culture: European Historical Perspectives*. Oxford: Berg Publishers, 71–86.

Leonard, Jane Kate. 1996. *Controlling from Afar: The Daoguang Emperor's Management of the Grand Canal Crisis, 1824–1826*. Ann Arbor, MI: Center for Chinese Studies, University of Michigan.

Levin, S. A. 1999. *Fragile Dominion: Complexity and the Commons*. Reading, MA: Perseus Books.

Lewis, D. L. 1987. *The Race to Fashoda: European Colonialism and the African Resistance in the Scramble for Africa*. New York: Weidenfeld & Nicolson.

Lewis, Samuel. 1848. *A Topographical Dictionary of England*, vol. 1, 7th ed. London: S. Lewis and Co.

Linant de Bellefonds, Louis Maurice Adolphe. 1872–3. *Mémoires sur les principaux travaux d'utilité publique exécutés en Egypte depuis la plus haute antiquité´ jusqu'à nos jours.* Paris: A. Bertrand.

Lindegren, Jan. 1988. *Varat, staten och diket: tre historieteoretiska uppsatser.* Uppsala, Sweden: Historiska institutionen.

Lindsay, Hugh Hamilton and Karl Friedrich A. Gützlaff. 1833. *Report of Proceedings on a Voyage to the Northern Ports of China, in the Ship Lord Amherst: Extracted from Papers, Printed by Order of the House of Commons, Relating to the Trade with China*, 2nd ed. London: B. Fellows.

Lindzen, Richard S. 1996. 'The importance and nature of the water vapor budget in nature and models', in H. LeTreut (ed.), *Climate Sensitivity Perturbations: Radiative Physical Mechanisms and Their Validation*. NATO ASI Series, I, 34: 51–66.

Linton, Jamie. *What Is Water?: The History of a Modern Abstraction*. Vancouver: UBC Press, 234.

Lipset, Seymour M. and Stein Rokkan. 1967. 'Cleavage structures, party systems, and voter alignments: An introduction', in Seymour M. Lipset and Stein Rokkan (eds), *Party Systems and Voter Alignments: Cross-National Perspectives*. New York: Free Press, 1–64.

Lipsey, Richard G. and Clifford Bekar. 2004. 'Science, institutions and the Industrial Revolution'. *Journal of European Economic History* 33.3: 709–53.

Litfin, Karen T. 1997. 'Sovereignty in world ecopolitics'. *Mershon International Studies Review* 41.2: 167–204.

Loáiciga, Hugo A. 2003. 'Climate change and ground water'. *Annals of the Association of American Geographers* 93.1: 30–41.

Locke, John. 1910. *Locke's Essay Concerning Human Understanding: Books II and IV* (with Omissions), ed. Mary Whiton. Chicago and London: Open Court Publishing.

Lombardini, E. 1865. *Essai sur l'hydrographie du Nil*. Paris: Challame.

Louis, William Roger (ed.). 1976. *Imperialism: The Robinson and Gallagher Controversy.* New York: New Viewpoints.

Louis, William Roger and Robin Winks (eds). 2001. *The Oxford History of the British Empire: Historiography*. Oxford: Oxford University Press.

Lovelock, James, 2000. Foreword, in Anne Primavesi, *Sacred Gaia*. New York: Routledge.

———. 2000. *Gaia: A New Look at Life on Earth*. Oxford: Oxford University Press.

Lugard, Frederick. D. 1893. *The Rise of Our East African Empire*, 2 vols. London: Blackwood.

Lunden, Kåre. 1981. 'Om årsakene til den norske bondefridomen', in Claus Krag and Jørn Sandnes, *Nye Middelalderstudier: Bosetning og Økonomi*. Oslo: Universitetsforlaget, 21–30.

———. 2002. *Norges landbrukshistorie II. 1350–1814. Frå svartedauden til 17. mai.* Oslo: Samlaget.

MacDonald, Murdoch. 1920. *Nile Control. A statement of the necessity for further control of the Nile to complete the development of Egypt and develop a certain area in the Sudan, with particulars of the physical conditions to be considered and a programme of the engineering works involved*, 2 vols, Cairo: Ministry of Public Works.

McGiffert, Arthur Cushman. 1897. *A History of Christianity in the Apostolic Age*. New York: Charles Scribner's Sons.

McKittrick, Meredith. 2006. 'The wealth of these nations: Rain, rulers and religion on the Cuvelai Floodplain', in Terje Tvedt and Terje Oestigaard (eds), *A History of Water: The World of Water*, Vol. III, Series I (series ed. Terje Tvedt). London and New York: I.B.Tauris, 449–70.

Macnaghten, Phil and John Urry. 1988. *Contested Natures*. London: Sage.

Macqueen, Norrie. 2011. *Humanitarian Intervention and the United Nations*. Edinburgh: Edinburgh University Press.

MacQuitty, William. 1976. *Island of Isis: Philae, Temple of the Nile*. London: Macdonald and Jane's.

Magnus, Sir Phillip Montefiore. 1958. *Kitchener, Portrait of an Imperialist*. London: Murray.

Major, David C. and Kenneth D. Frederick. 1997. 'Water resources planning and climate change assessment methods'. *Climatic Change* 37.1: 25–40.

Mantoux, Paul. 1970 [1928]. *The Industrial Revolution in the Eighteenth Century: An Outline of the Beginnings of the Modern Factory System in England*. London: Methuen.

Marks, Robert B. 2002. *The Origins of the Modern World: A Global and Ecological Narrative*. Lanham, MD: Rowman & Littlefield.

Marriott, McKim and Ronald B. Inden, 1977. 'Toward an ethnosociology of South Asian caste systems', in Kenneth David (ed.), *The New Wind: Changing Identities in South Asia*. The Hague and Paris: Mouton, 227–38.

Martin, Lisa. 1994. 'Heterogeneity, linkage and common problems'. *Journal of Theoretical Politics* 6.4: 473–93.

Marx, Karl. 1959 [1844]. *Economic and Philosophic Manuscripts of 1844*. Moscow: Progress Publishers. Available at https://www.marxists.org/archive/marx/works/1844/manuscripts/preface.htm. (Accessed 24 June 2014.)

———. 1867 [1906]. *Capital, Capital: A Critique of Political Economy, Vol. I. The Process of Capitalist Production*. Chicago: Charles H. Kerr and Co. Available at https://www.marxists.org/archive/marx/works/1867–c1/ch16.htm. (Accessed 24 June 2014.)

Mason Bey. 1881. 'Note sur les nilomêtres et le mesurage des affluents du Nil, notamment du Nil blanc'. *Bulletin de Société de geographie d'Egypte* 1–2: 51–6.

Mayor, Adrienne. 2011. *The First Fossil Hunters: Paleontology in Greek and Roman Times: With a New Introduction by the Author*. Princeton, NJ: Princeton University Press.

Mellor, Roy E. H. 1983. *The Rhine: A Study in the Geography of Water Transport*. O'Dell Memorial Monograph 16. Aberdeen: Department of Geography.

Melosi, M. 1993. 'The place of the city in environmental history'. *Environmental History Review* 17.1: 1–23.

———. 2008. *The Sanitary City: Environmental Services in Urban America from Colonial Times to the Present*. Pittsburgh, PA: University of Pittsburg Press.

Michel, Aloys A. 1967. *The Indus Rivers: A Study of the Effects of Partition*. New Haven, CT: Yale University Press.

Milner, Alfred. 1892. *England in Egypt*. London: E. Milner.

Milton, A. S. 1998. 'Thermal physiology: Brief history and perspectives', in C. M. Blatteis (ed.), *Physiology and Pathophysiology of Temperature Regulation*. Singapore: World Scientific Publishing, 3–11.

Morimoto, T. 1998. 'Heat loss mechanisms', in C. M. Blatteis (ed.), *Physiology and Pathophysiology of Temperature Regulation*. Singapore: World Scientific Publishing, 80–91.

Morrissette, Jason and Douglas A. Borer. 2004. 'Where oil and water do mix: Environmental scarcity and future conflict in the Middle East and North Africa'. *Parameters* 34(4): 86–101.

Morse Edward. 1976. *Modernization and the Transformation of International Relations*. New York: Free Press.

Motzfeldt, U. A. 1908. *Den norske Vasdragsrets Historie indtil Aaret 1800*. Kristiania, Denmark: Brøgger i Comm.

Mouri, Goro. 2014. 'An academic goal of socio-ecological sustainability: A comprehensive review from a millennial-scale perspective'. *International Journal of Sustainable Built Environment* 3(1): 47–53.

Mwiandi, Mary C. 2010. 'The Nile waters and the socio-economic development of western Kenya', in Terje Tvedt (ed.), *The River Nile in the Post-Colonial Age: Conflict and Cooperation Among the Nile Basin Countries*. London: I.B.Tauris, 93–25.

Mykland, Knut. 1976. *Norges Historie*. Oslo: Cappelen Forlag.

Nalbantis, I., D. Koutsoyiannis and Th Xanthopoulos. 1992. 'Modeling the Athens water supply system'. *Water Resources Management* 6.1: 57–67.

Namafe, Charles. 2006. 'The Lozi flood tradition', in Terje Tvedt and Terje Oestigaard (eds), *A History of Water: The World of Water*, Vol. III, Series I (series ed. Terje Tvedt). London and New York: I.B.Tauris, 470–88.

Narayan, Rajdeva and Janardan Kumar. 2003. *Ecology and Religion: Ecological Concepts in Hinduism, Buddhism, Jainism, Islam, Christianity and Sikhism*. New Delhi: Deep & Deep Publications.

Needham, Joseph. 1996. *Science and Civilization in China*. New York: Cambridge University Press.

Ngowi, Honest Prosper. 2010. 'Unlocking economic growth and development potential in sleeping giants: the Nile Basin approach in the case of Tanzania', in Terje Tvedt (ed.), *The River Nile in the Post-Colonial Age: Conflict and Cooperation Among the Nile Basin Countries*. London: I.B.Tauris, 57–73.

Nielsen, B. 1998. 'Temperature regulation in exercise', in C. M. Blatteis (ed.), *Physiology and Pathophysiology of Temperature Regulation*. Singapore: World Scientific Publishing, 128–43.

Nijssen, Bart, Greg M. O'Donnell, Alan F. Hamlet, et al. 2001. 'Hydrologic sensitivity of global rivers to climate change'. *Climatic Change* 50.1–2:143–75.

Nishijima, Sadao. 1984. 'The formation of the early Chinese cotton industry', in Linda Grove and Christian Daniels (eds), *State and Society in China: Japanese Perspectives on Ming-Quing Social and Economic History*. Tokyo: University of Tokyo Press, 17–79.

Norberg, J. and G. S. Cumming. 2008. *Complexity Theory for a Sustainable Future*. New York: Columbia University Press.

Nordström, Byron J. (ed.). 1986. *Dictionary of Scandinavian History*. London: Greenwood Press.

Nordtveit, Ernst. 2015. 'History of Scandinavian water law', in Terje Tvedt, Owen McIntyre and Tadess Kasse Woldetsadik (eds), *A History of Water: Sovereignty and International Water Law*, Vol. 2, Series III (series ed. Terje Tvedt). London and New York: I.B.Tauris, 105–127.

North, Douglass C. 1981. *Structure and Change in Economic History*. New York: W. W. Norton.

North, Douglass C. and Robert Paul Thomas. 1973. *The Rise of the Western World: A New Economic History*. New York: Cambridge University Press.

Nunes Correia, F. and J. da Silva. 1999. 'International framework for the management of transboundary water resources'. *Water International* 24.2: 86–94.

O'Brien, Patrick K. 1987. *Economic Growth in Britain and France, 1780–1914: Two Paths to the Twentieth Century*. London: Allen and Unwin.

Oestigaard, Terje. 2005. *Death and Life-giving Waters: Cremation, Caste, and Cosmogony in Karmic Traditions*. BAR International Series 1353, Oxford.

———. 2006. 'Purification, purgation and penalty: Christian concepts of water and fire in Heaven and Hell', in Terje Tvedt and Terje Oestigaard (eds), *A History of Water: Ideas of Water from Ancient Societies to the Modern World*, Vol. III, Series I (series ed. Terje Tvedt). London and New York: I.B.Tauris, 298–323.

———. 2013. *Water, Christianity and the Rise of Capitalism*. London: I.B.Tauris.

Oki, Taikan and Shinjiro Kanae. 2006. 'Global hydrological cycles and world water resources'. *Science* 313.5790: 1068–72.

Oliver, John E. 1973. *Climate and Man's Environment*. New York: John Wiley.

O'Neill, John. 1993. *Ecology, Policy and Politics: Human Well-Being and the Natural World*. London and New York: Routledge.

Ophuls, William. 1977. *Ecology and the Politics of Scarcity: Prologue to a Political Theory of the Steady State*. San Francisco: W. H. Freeman.

Opie, John. 1993. *Ogallala: Water for a Dry Land*. Lincoln: University of Nebraska Press.

Orum, Anthony M. and Xiangming Chen. 2003. *The World of Cities. Places in Comparative and Historical Perspective*. Series on 21st-Century Sociology. London: Blackwell.

Ostrom, Elinor, et al. (eds). 2002. *The Drama of the Commons*. Washington, DC: National Academy Press.

Owen, Roger. 1981. *The Middle East in the World Economy, 1800–1914*. London: Methuen.

Pacey, Arnold. 1991. *Technology in World Civilization: A Thousand-Year History*. Cambridge, MA: MIT Press.

Pacione, Michael. 2009. *Urban Geography: A Global Perspective*. London: Routledge.

Pakenham, Thomas. 1992. *The Scramble for Africa, 1876–1912*. London: Abacus.

Parker, Simon. 2003. *Urban Theory and the Urban Experience: Encountering the City*. London: Routledge.

Parry, Jonathan. 1985. 'Death and digestion: The symbolism of food and eating in North Indian mortuary rites'. *Man* 20.4: 612–30.

———. 1994. *Death in Banaras*. The Lewis Henry Morgan Lectures 1988. Cambridge: Cambridge University Press.

Parthasarathi, Prasannan. 2001. *The Transition to a Colonial Economy: Weavers, Merchants and Kings in South-India, 1720–1800*. Cambridge: Cambridge University Press.

Peel, Sidney Cornwallis. 1969 [1904]. *The Binding of the Nile and the New Soudan*. New York: Negro Universities Press.

Pepper, David. 1996. *Modern Environmentalism: An Introduction*. London: Routledge.

Pepper, David, Frank Webster and George Revill. 2003. *Environmentalism: Critical Concepts*, 5 vols. London: Routledge.

Perkin, Harold James. 1969. *The Origins of Modern English Society 1780–1880*. London: Routledge & Kegan Paul.

Petts, Geoffrey E. and C. Amoros (eds). 1996. *Fluvial Hydrosystems*. London: Chapman & Hall.

Phillips, J. 1803. *A General History of Inland Navigation, Foreign and Domestic: Containing a Complete Account of the Canals already Executed in England with Consideration of those Projects*, 4th ed. London: J. Taylor.

Picaut, J., F. Masia and Y. du Penhoat. 1997. 'An advective-reflective conceptual model for the oscillatory nature of the ENSO'. *Science* 277.5326: 663–6.

Platt, Harold. L. 2005. *Shock Cities: The Environmental Transformation and Reform of Manchester and Chicago*. Chicago: University of Chicago Press.

Polanyi, Karl. 1944. *The Great Transformation*. Boston, MA: Beacon Press.

Pollard, Sidney. 1964. 'Fixed capital in the Industrial Revolution in Britain'. *Journal of Economic History* 24.3: 120–41.

Pomeranz, Kenneth. 2000. *The Great Divergence: China, Europe and the Making of the Modern World Economy*. Princeton, NJ: Princeton University Press.

Possehl, Gregory L. 2010. 'The Indus civilization and riverine history in Northwestern India and Pakistan', in Terje Tvedt and Richard Coopey (eds), *A History of Water: Rivers and Society: From the Birth of Agriculture to Modern Times*, Vol. II, Series II (series ed. Terje Tvedt), London and New York: I.B.Tauris, 29–52.

Potter, Thomas D. and Bradley R. Colman (eds). 2003. *Handbook of Weather, Climate and Water*. New York: John Wiley.

Pounds, Norman J. G. 1973. *An Historical Geography of Europe 450 B.C.–A.D. 1330*. Cambridge: Cambridge University Press.

Pratt, Vernon, Jane Howarth and Emily Brady. 2000. *Environment and Philosophy*. London: Routledge.

Proctor, James D. 1998. 'The social construction of nature: Relativist accusations, pragmatist and critical realist responses'. *Annals of the Association of American Geographers* 88.3: 352–76.

———. 2001. 'Concepts of nature, environmental/ecological', in N. J. Smelser and P. B. Bates (eds), *International Encyclopedia of the Social and Behavioral Sciences*. Oxford: Elsevier Science, 10400–6.

Prompt, Victor. 1893. *Soudan Nilotique*, 20 January 1893.

Qin, Boqiang, and Qun Huang. 1998. 'Evaluation of the climatic change impacts on the inland lake: A case study of Lake Qinghai, China'. *Climatic Change* 39.4: 695–714.

Radkau, Joachim. 2008. *Nature and Power: A Global History of the Environment*. Cambridge: Cambridge University Press.

Raychaudhuri, Tapan. 1982. 'Non-agricultural production, 1. Moghul India', in Tapan Raychaudhuri and Irfan Habib (eds), *The Cambridge Economic History of India*, Vol. 1, *c.1200–c.1750*. Cambridge: Cambridge University Press, 261–308.

Redford, Arthur. 1960. *The Economic History of England 1760–1860*. Westport, CT: Greenwood Press.

Reynolds, Terry S. 1983. *Stronger Than a Hundred Men: A History of the Vertical Water Wheel*. Baltimore, MD: Johns Hopkins University Press.

Rind, D., C. Rosenzweig and R. Goldberg. 1992. 'Modelling the hydrological cycle in assessments of climate change'. *Nature* 358: 119–23.

Rinne, Katherine. 2011. *The Waters of Rome: Aqueducts, Fountains and the Birth of the Baroque City*. New Haven, CT: Yale University Press.

———. 2014. 'Plumbing Ancient Rome', in Terje Tvedt and Terje Oestigaard (eds), *A History of Water: Water and Urbanization*, Vol. I, Series III, London and New York: I.B.Tauris, 149–72.

Rivlin, Helen Anne B. 1961. *The Agricultural Policy of Muhammad 'Ali in Egypt*. Cambridge, MA: Harvard University Press.

Robinson, Ronald. 1959. 'Imperial problems in British politics, 1880–1895', in E. A. Benians, J. Butler and C. E. Carrington (eds), *The Cambridge History of Empire*, vol. 3. The Empire Commonwealth. Cambridge: Cambridge University Press, 127–80.

Robinson, Ronald and John Gallagher. 1961. *Africa and the Victorians: The Climax of Imperialism in the Dark Continent*. New York: St. Martins Press.

Robinson, Ronald, John Gallagher and Alice Denny. 1981. *Africa and the Victorians: The Official Mind of Imperialism*. London: Macmillan.

Robson, Robert. 1957. *The Cotton Industry in Britain*. London: Macmillan.

Roehl, Richard. 1976. 'French industrialization: A reconsideration'. *Explorations in Economic History* 13.3: 233–81.

Rolston, Holmes. 1992. 'Challenges in environmental ethics', in David F. Cooper and Joy A. Palmer (eds), *The Environment in Question*. London: Routledge, 135–46.

Ronan, Colin A. 1995. *The Shorter Science and Civilization in China*, Vol. 5. Cambridge: Cambridge University Press.

Rosecrance, Richard. 1986. *The Rise of the Trading State: Commerce and Conquest in the Modern World*. New York: Basic Books.

Rosen, Christine M. and Joel Arthur Tarr. 1994. 'The importance of an urban perspective in environmental history'. *Journal of Urban History* 20: 299–310.

Ross, J. C. P. 1893. 'Irrigation and agriculture in Egypt'. *Scottish Geographical Magazine* 9.4: 161–93.

Rostow, Walter W. 1960. *Stages of Economic Growth: A Non-Communist Manifesto*. Cambridge: Cambridge University Press.

Rudolph, Christopher. 2005. 'Sovereignty and territorial borders in a global age'. *International Studies Review* 7.1: 1–20.

Ruiz, José Luis Martínez and Daniel Murillo Licea. 2010. 'Water in the cosmovision and symbolism of Mesoamerica and Peru in the Pre-Hispanic Period', in Terje Tvedt and Terje Oestigaard (eds), *A History of Water: Ideas of Water From Ancient Societies to the Modern World*, Vol. I, Series II, London and New York: I.B.Tauris, 440–73.

Russell, Brian. 2001. *Baptism, Sign and Seal of the God's Grace*. London: Grace Publications Trust.

Rzoska, Julian (ed.). 1976. *The Nile: Biology of an Ancient River*. The Hague: Junk.

Salby, Murray L. 1996. *Fundamentals of Athmospheric Physics*. Salt Lake City, UT: Academic Press.

Sanderson, George Neville. 1965. *England, Europe and the Upper Nile 1882–1899*. Edinburgh: Edinburgh University Press.

Sanderson, Neville and Lilian. P. Sanderson. 1981. *Education, Religion and Politics in Southern Sudan 1899–1964*. Sudan Studies, vol. 4. London: Ithaca Press.

Santamaria, Ulysses and Anne M. Bailey. 1984. 'A note on Braudel's structure as duration'. *History and Theory* 23.1: 78–83.

Sarachik, Edward S. 2003. 'The ocean in climate', in Thomas D. Potter and Bradley R. Colman (eds), *Handbook of Weather, Climate, and Water: Dynamics, Climate, Physical Meteorology, Weather Systems, and Measurements*. Hoboken, NJ: John Wiley, 129–35.

Schaefer, F. K. 1953. 'Exceptionalism in geography: A methodological examination'. *Annals of the Association of American Geographers* 43.3: 226–49.

Scott-Moncrieff, C. 1895. 'The Nile'. *Royal Institution of Great Britain, Proceedings* 14.25 (January): 405–18.

Selin, H. and A. Kalland (eds). 2003. *Nature Across Cultures: Views of Nature and the Environment in Non-Western Cultures*. Dordrecht, Netherlands: Kluwer Academic Publishers.

Sivapalan, M. and J. D. Kalma. 1995. 'Scale problems in hydrology', in J. D. Kalma and M. Sivapalan (eds), *Scale Issues in Hydrological Modelling*. Chichester: Wiley, 1–8.

Smith, Simon. 1998. *British Imperialism, 1750–1970*. Cambridge: Cambridge University Press.

Sharma, Arvind and Mahdu Kanna. 2013. *Asian Perspectives on the World's Religions after September 11*. Westport, CT: Praeger.

Shaw, J. 2006. 'Landscape, water and religion in ancient India', *International Archaeology* 9: 43–8.

Shipley, A. E. 1936 [1925]. *Life: A Book for Elementary Students*. Cambridge: Cambridge University Press.

Shiue, Carol H. and Wolfgang Keller. 2004. 'Markets in China and Europe on the eve of the Industrial Revolution'. Working paper 10778, *National Bureau of Economic Research*, September 2004.

Simmons, I. G. 1993. *Environmental History: A Concise Introduction*. Oxford: Oxford University Press.

Smith, Adam. 1937 [1776]. *The Wealth of Nations*. New York: Modern Library.

Smith, Herbert Arthur. 1931. *The Economic Use of International Rivers*. London: P. S. King.

Soja, Edward. 1989. *Postmodern Geographies: The Reassertion of Space in Critical Theory*. London: Verso.

Solomon, Steven. 2010. *Water. The Epic Struggle for Wealth, Power, and Civilization*. New York: Harper Collins.

Sprout, Harold and Margaret Sprout. 1962. *Foundation of International Politics*. Princeton, NJ: Princeton University Press.

Starr, Joyce. R. 1991. 'Water wars'. *Foreign Policy* 82: 17–36.

Stefan, H. G., X. Fang and M. Hondzo. 1998. 'Simulated climate change effects on year-round water temperatures in temperate zone lakes'. *Climatic Change* 40.3–4: 547–76.

Steffen, W. 2001. 'Toward a new approach to climate impact studies', in L. Bengtsson and C. Hammer (eds), *Geosphere-Biosphere Interactions and Climate*. Cambridge: Cambridge University Press, 273–9.

Steinsland, Gro. 2005. *Norrøn Religion: Myter, Riter, Samfunn*. Oslo: Pax.

Steward, Julian Haynes. 1972. *Theory of Culture Change: The Methodology of Multilinear Evolution*. Urbana, IL: University of Illinois Press.

Steward, Julian (ed.). 1972. *Contemporary Change in Traditional Societies*. A Special Publication of the Illinois Studies in Anthropology, 3 vols. Urbana: University of Illinois Press.

Strand, Sverre, 1973. *Urban Geography, 1950–'70. A Comprehensive Bibliography of Urbanism as Reflected in the Articles and Book Reviews of 72 American, Canadian, British, Dutch, and Scandinavian Geographical Periodicals*. Monticello, IL: Council of Planning Librarians.

Strang, Veronica. 2010. 'Water in Aboriginal Australia', in Terje Tvedt and Terje Oestigaard (eds), *A History of Water: Ideas of Water From Ancient Societies to the Modern World*, Vol. II, Series II. London and New York: I.B.Tauris, 343–78.

Stunkel, Kenneth R. 2011. *Fifty Key Works of History and Historiography*. London: Routledge.

Swingewood, A. 1991. *A Short History of Sociological Thought*. London: Macmillan Press.

Swyngedouw, Eric. 1996. 'The city as a hybrid: On nature, society and cyborg urbanisation'. *Capitalism, Nature, Socialism* 7.1: 65–80.

———. 1997. 'Power, nature and the city. The conquest of water and the political ecology of urbanization in Guayaquil, Ecuador, 1880–1990'. *Environment and Planning* A 29: 311–32.

———. 1999. 'Modernity and hybridity: Nature, *regeneracionismo*, and the production of the Spanish waterscape, 1890–1930'. *Annales of the Association of American Geographers* 89.3: 443–65.

———. 2004. *Social Power and the Urbanization of Water: Flows of Power*. Oxford: Oxford University Press.

———. 2006. 'Circulation and metabolisms: (Hybrid) natures and (cyborg) cities'. *Science as Culture* 15: 105–21.

Szostak, Rick. 1991. *The Role of Transportation in the Industrial Revolution*. Montreal: McGill-Queen's University Press..

Sætersdal, Tore. 2010. 'Rain, snakes and sex – making rain: rock art and rain-making in Africa and America', in Terje Tvedt and Terje Oestigaard (eds), *A History of Water: Ideas of Water from Ancient Societies to the Modern World*, Vol. I, Series II (series ed. Terje Tvedt). London and New York: I.B.Tauris, 378–405.

The Articles of the Treaty of Westphalia. Peace Treaty signed and sealed at Munster in Westphalia the 24th October, 1648. 1697. London: W. Onley.

Tignor, Robert. L., 1966. *Modernization and British Colonial Rule in Egypt, 1882–1914*. Princeton, NJ: Princeton University Press.

Tottenham, P. M. 1926. *Upper White Nile Mission. Interim Report 1923*. Cairo: Government Press.

Toynbee, Arnold. 1984 [1884]. *The Industrial Revolution*. London: Green.

Treaty of Westphalia. Peace Treaty between the Holy Roman Emperor and the King of France and Their Respective Allies. Available at: http://avalon.law.yale.edu/17th_century/westphal.asp. (Accessed 13 June 2014.)

Trenberth, Kevin. E., John T. Fasullo and Jeffrey Kiehl. 2009. 'Earth's global energy budget'. *Bulletin of the American Meteorological Society* 90: 311–23.

Tucker, Mary Evelyn and John Grim. 1994. *Worldviews and Ecology: Religion, Philosophy, and the Environment*. Maryknoll, NY: Orbis Books.

Turnbull, Gerard. 1987. 'Canals, coal and regional growth during the Industrial Revolution'. *Economic History Review* 40.4: 537–60.

Turner, N. J. 2005. *The Earth's Blanket: Traditional Teachings for Sustainable Living*. Seattle, WA: University of Washington Press.

Tvedt, Terje. 1997. *En reise i vannets historie*. Oslo: Cappelen.

———. 2004a. *The River Nile in the Age of the British: Political Ecology and the Quest for Economic Power*. London and New York: I.B.Tauris.

———. 2004b. *The Nile. An Annotated Bibliography*, 2nd ed. London and New York: I.B.Tauris.

281

———. 2004c. *The Southern Sudan. An Annotated Bibliography*, 2 vols, 2nd ed. London and New York: I.B.Tauris.

———. 2007. *En Reise i Vannets Fremtid*. Oslo: Kagge Forlag.

———. 2012. *Nilen. Historiens Elv*. Oslo: Aschehoug.

———. 2014. *A Journey in the Future of Water*. London: I.B.Tauris.

——— (ed.). 2009. *The River Nile in the Post-Colonial Age: Conflict and Cooperation Among the Nile Basin Countries*. London: I.B.Tauris.

Tvedt, Terje and Terje Oestigaard (eds). 2006. *A History of Water: Worlds of Water*, Vol. III, Series I. London and New York: I.B.Tauris.

——— (eds). 2010. *A History of Water: Ideas of Water*, Vol. I, Series II. London and New York: I.B.Tauris.

——— (eds). 2014. *A History of Water: Water and Urbanization*, Vol. I, Series III. London and New York: I.B.Tauris.

Tvedt, Terje, Graham Chapman and Roar Hagen (eds). 2010. *A History of Water: The Geopolitics of Water*, Vol. III, Series II. London and New York: I.B.Tauris.

Tvedt, Terje, Owen Roberts and Tadesse Kassa (eds). 2014. *A History of Water: Sovereignty and International Water Law*, Vol II, Series III. London and New York: I.B.Tauris.

Unwin, George, Arthur Hulme and George Taylor. 1924. *Samuel Oldknow and the Arkwrights: The Industrial Revolution at Stockport and Marple*. Manchester: Manchester University Press.

Van Huyssteen, J. Wentzel. 1999. *The Shaping of Rationality: Toward Interdisciplinarity in Theology and Science*. Grand Rapids, MI: W. B. Eerdmans.

Vatikiotis, P. J. 1991. *The History of Modern Egypt, from Muhammad Ali to Mubarak*. Baltimore: Johns Hopkins University Press. (lst ed. 1976, London: Weidenfeld and Nicholson.)

Ven, G. P. van de. 1993. *Man-Made Lowlands: History of Water Management and Land Reclamation in the Netherlands*. Utrecht: Uitgeverij Matrijs.

Ventre Bey, F. 1893. *Hydrologie du Bassin du Nil: Essai sur la prévision des crues du fleuve*. Cairo: Imp. Nationale.

Verghese, B. G. 1990. *Waters of Hope: Integrated Water Resource Development and Regional Cooperation within the Himalayan-Ganga-Brahmaputra-Barak Basin*. Dhaka: Academic Publishers.

Von Tunzelmann, G. N. 1978. *Steam Power and British Industrialization to 1860*. Oxford: Clarendon Press.

Vorosmarty, Charles, P. Green, J. Salisbury et al. 2000. 'Global water resources: Vulnerability from climate change and population growth'. *Science* 289.5477: 284–8.

Wagner, Donald B. 1984. 'Some traditional Chinese iron production techniques practiced in the 20th century'. *Journal of the Historical Metallurgy Society* 18.2: 95–104.

Wallerstein, Immanuel. 1998. 'Time and duration: The unexcluded middle, or reflections on Braudel and Prigogine'. *Thesis Eleven* 54: 79–87.

———. 1998. 'The time of space and the space of time: the future of social sciences'. *Political Geography* 17.1: 71–82.

Wallis, Michael T., Michael R. Ambrose and Clifford C. Chan. 2008. 'Climate change: Charting a water course in an uncertain future'. *Journal of the American Water Works Association* 100.6: 70–9.

Ward, Paul W. 1928. *Sovereignty. A Study of a Contemporary Political Notion*. London: George Routledge.

Ward, Roy C. and Mark Robinson. 2000. *Principles of Hydrology*. London: McGraw-Hill.

Warren, D. 1995. *With Broadax and Firebrand: The Destruction of the Brazilian Atlantic Forest*. Berkeley, CA: University of California Press.

Watts, Martin. 2005. *Water and Wind Power*. Aylesbury, UK: Shire Publications.

Weber, Max. 1930 [1905]. *The Protestant Ethic and the Spirit of Capitalism*, trans. Talcott Parsons. New York: Charles Scribner's Sons.

———. 1946. *From Max Weber: Essays in Sociology*, trans. and ed. H. H. Gerth and C. Wright Mills. New York: Oxford University Press.

———. 1949. *Max Weber on the Methodology of the Social Sciences*, ed. Edward Shils and Henry A. Finch. Glencoe, IL: Free Press.

———. 1958. *The City*, trans. and ed. R, Martindale and G. Neuwirth. New York: The Free Press.

———. 1963 [1920]. *The Sociology of Religion*. London: Methuen.

———. 1968 [1922]. *Economy and Society: An Outline of Interpretive Sociology*, ed. G. Roth and C. Wittich, 3 vols. New York: Bedminster Press.

———. 1991 [1922]. 'The nature of social action', in W. G. Runciman (ed.), *Weber: Selections in Translation*. Cambridge: Cambridge University Press, 7–32.

Weber, Max and Wolf V. Heydebrand. 1994. *Sociological Writings*. London and New York: Continuum.

Webster, Anthony. 2006. *The Debate on the Rise of the British Empire*. Manchester: Manchester University Press.

Weiner, Ronald R. 2005. *Lake Effects: A History of Urban Policy Making in Cleveland, 1825–1929*. Columbus: Ohio State University Press.

Westra, Christian. 2009. 'Will the "Bush Doctrine" survive its progenitor? An assessment of jus ad bellum norms for the post-Westphalian Age'. *Boston College International and Comparative Law Review* 32.2: 398–422.

Wheeler, Tim and Joachim von Braun. 2013. 'Climate change impacts on global food security'. *Science* 341.6145: 508–13.

Whitbeck, Ray H. and Olive J. Thomas. 1932. *The Geographic Factor*. New York: Appleton-Century-Crofts.

White, Arthur Silva. 1899. *The Expansion of Egypt under Anglo-Egyptian Condominium*. London: Methuen.

White, Lynn. Jr. 1967. 'The historical roots of our ecological crisis'. *Science* 155.3767: 1203–7.

White, Richard, 1996. *The Organic Machine. The Remaking of the Columbia River*. New York: Hill and Wang.

Whitehand, J. W. R. 1977. 'The basis for an historico-geographical theory of urban form'. *Transactions of the Institute of British Geographers* 2: 400–16.

WHO. 2013. *The Global Climate 2001–2010: A Decade of Climate Extremes. Summary Report*. Geneva: WHO.

Wiens, Herold J. 1955. 'Riverine and coastal junks in China's commerce'. *Economic Geography* 31.3: 248–64.

Willan, Thomas Stuart. 1964. *River Navigation in England, 1600–1750*. London: F. Cass.

Willcocks, William. 1889. *Egyptian Irrigation*, 2 vols. New York: E. & F. N. Spoon.

———. 1893. *Report on the Nile and Proposed Reservoirs*. Cairint 3/14/232, NRO, Khartoum, Sudan.

———. 1894. *Report on Perennial Irrigation and Flood Protection of Egypt*. Cairo: Ministry of Public Works.

———. 1936. *Sixty Years in the East*. London: E. & F. N. Spoon.

Willcocks, William and James Craig. 1913. *Egyptian Irrigation*, 2nd ed. London: E. & F. N. Spoon.

Willey, Gordon R. and Jeremy A. Sabloff. 1980. *A History of American Archaeology*. San Francisco: W. H. Freeman.

Williams, Daniel and Susan Stewart. 1998. 'Sense of place: An elusive concept that is finding a home in ecosystem management'. *Journal of Forestry* 96.5: 18–23.

Williams, Eric. 1961 [1944]. *Capitalism and Slavery*. New York: Russel and Russel.

Williams, R. 1988. *Keywords: A Vocabulary of Culture and Society*. London: Fontana.

Williams, Raymond. 1980. *Problems in Materialism and Culture: Selected Essays*. London: Verso.

283

Williamson, Jeffrey G. and Laura Panza. 2013. 'Did Muhammad Ali foster industrialization in early 19th century Egypt?' CEPR Discussion Paper, No. DP9363.

Willmott, W. E. (ed.). 1972. *Economic Organization in Chinese Society*. Stanford, CA: Stanford University Press.

Winks, Robin W. 2001. *The Oxford History of the British Empire*, Vol. 5. Oxford: Oxford University Press.

Wittfogel, Karl. 1981 [1957]. *Oriental Despotism: A Comparative Study of Total Power*. New Haven, CT: Yale University Press.

Wolf, Eric R. 1964 *Anthropology: Humanistic Scholarship in America*. New Jersey: Prentice-Hall.

Wood, W. H. Arden. 1924. 'Rivers and man in the Indus-Ganges alluvial plain'. *Scottish Geographical Magazine* 40.1: 1–16.

Woolf, Daniel. 2011. *A Global History of History*. Cambridge: Cambridge University Press.

World Commission of Environment and Development. 1987. *Our Common Future*. Oxford: Oxford Paperbacks.

World Water Assessment Programme. 2003. 'Water for people, water for life: The United Nations World Water Development Report'. Paris: UNESCO.

Worster, Donald. 1985a [1977]. *Nature's Economy: A History of Ecological Ideas*. New York and Cambridge: Cambridge University Press.

———. 1985b. *Rivers of Empire: Water, Aridity and the Growth of the American West*. New York: Pantheon books.

——— (ed.). 1988. *The Ends of Earth: Perspectives on Modern Environmental History*. Cambridge: Cambridge University Press.

———. 1990. 'Transformations of the Earth: Towards an agroecological perspective in history'. *Journal of American History* 74.4: 1087–106.

Worthington, E. Barton. 1929. *A Report on the Fishing Survey of Lakes Albert and Kioga. March to July, 1928. With Appendices I to V, 2 Maps and 24 Other Illustrations* [London]: pub. on behalf of the government of the Uganda protectorate by the crown agents for the colonies.

Wunderlich, Jens-Uwe and Meera Warrie. 2010. *A Dictionary of Globalization*. London: Routledge.

Wynne-Jones, Stephania and Jeffrey Fleisher. 2014. 'Water management in a maritime culture: the Swahili coast of East Africa', in Terje Tvedt and Terje Oestigaard (eds), *A History of Water: Water and Urbanization*, Vol. I, Series III. London and New York: I.B.Tauris, 242–62.

Yu, Liansheng. 2002. 'The Huanghe (Yellow) River: A review of its development, characteristics, and future management issues'. *Continental Shelf Research* 22.3: 389–403.

Zetland, Marquis of. 1932. *Lord Crome, being the Authorized Life of Evelyn Baring, Earl of Cromer*. London: Hodder and Stoughton.

Zhang, Qiang, Chong-Yu Xu, Tao Yang, et al. 2010. 'The historical developments and anthropogenic influences of the Yellow River up to the nineteenth century', in Terje Tvedt and Richard Coopey (eds), *A History of Water: Rivers and Society from the Birth of Agriculture to Modern Times*, Vol. II, Series II (series ed. Terje Tvedt). London and New York: I.B.Tauris, 144–65.

Zheng Xaio Yun. 2010. 'Shaping beliefs, identities and institutions: The role of water myths among ethnic groups in Yunnan, China', in Terje Tvedt and Terje Oestigaard (eds), *A History of Water: Ideas of Water from Ancient Societies to the Modern World*, Vol. II, Series II (series ed. Terje Tvedt). London and New York: I.B.Tauris, 405–25.

Zijderveld, Anton C. 2009. *Theory of Urbanity: The Economic and Civil Culture of Cities*. London: New Brunswick.

INDEX